AF183941

CIGRE Green Books

Series editor

Cigré, International Council on Large Electric Systems, Paris, France

CIGRE presents their expertise in compact professional books on electrical power networks. These books are of a self-contained concise character, covering the entire knowledge of the subject within power engineering. The books are created by CIGRE experts within their study committees and are recognized by the engineering community as the top reference books in their fields.

More information about this series at http://www.springer.com/series/15383

Carlos Samitier
Editor

Utility Communication Networks and Services

Specification, Deployment and Operation

cigré

Springer

Editor
Carlos Samitier
Pullnet Technology
Parc Technològic BC Nord
Barcelona
Spain

Co-editor
Mehrdad Mesbah
GE Energy Connections
Paris
France

ISSN 2367-2625 ISSN 2367-2633 (electronic)
CIGRE Green Books
ISSN 2509-2812 ISSN 2509-2820 (electronic)
Compact Studies
ISBN 978-3-319-82068-2 ISBN 978-3-319-40283-3 (eBook)
DOI 10.1007/978-3-319-40283-3

© Springer International Publishing Switzerland 2017
Softcover reprint of the hardcover 1st edition 2016
This work is subject to copyright. All rights are reserved by the Publisher, whether the whole or part of the material is concerned, specifically the rights of translation, reprinting, reuse of illustrations, recitation, broadcasting, reproduction on microfilms or in any other physical way, and transmission or information storage and retrieval, electronic adaptation, computer software, or by similar or dissimilar methodology now known or hereafter developed.
The use of general descriptive names, registered names, trademarks, service marks, etc. in this publication does not imply, even in the absence of a specific statement, that such names are exempt from the relevant protective laws and regulations and therefore free for general use.
The publisher, the authors and the editors are safe to assume that the advice and information in this book are believed to be true and accurate at the date of publication. Neither the publisher nor the authors or the editors give a warranty, express or implied, with respect to the material contained herein or for any errors or omissions that may have been made.

Printed on acid-free paper

This Springer imprint is published by Springer Nature
The registered company is Springer International Publishing AG Switzerland

Message from the Secretary General

All CIGRE members have access to the publications produced by the Working Groups of our Association, in the form of "Technical Brochures" when their work is completed.

Between 40 and 50 new Technical Brochures are published yearly, and these brochures are announced in Electra, CIGRE's bimonthly journal, and are available for downloading from "e-cigre". This online library of CIGRE (http://www.e-cigre.org), is one of the most comprehensive accessible databases of relevant technical works on power engineering.

During 1931 to 1974 the technical reports of Working Groups were published in Electra only. As some reports were becoming voluminous it was decided to publish the largest ones separately, as Technical Brochures.

In 2000, Electra was redesigned, and as a result, it no longer published the final reports of Working Groups. Today only summaries of Technical Brochures are provided in Electra, in both English and French.

The idea to value the collective work of the Study Committees accumulated over more than twenty years, by putting together all the Technical Brochures of a given field, in a single book, which was first proposed by Dr. Konstantin Papailiou to the Technical Committee in 2011.

As a result, the first two Green Books were published by CIGRE at the time of the last 2014 session in Paris. It is planned that the series will grow progressively at a pace of about one Green Book per year.

In 2015, the Technical Committee decided to associate CIGRE with a renowned publisher of scientific books in order to benefit from its know-how and of its visibility worldwide. An agreement was signed with Springer at the end of this year to co-publish the future CIGRE Green Books.

This new Green Book is the first result of this new collaboration, and it is interesting to note, that given its size, not as big as the first two books, together with our partner we decided to have an alternative format, the CIGRE Compact Studies, to satisfy the need of some Study Committees for a more concise book format covering a specific subject from the scope of a Study Committee.

Before closing this foreword, I want to thank and congratulate the editor, the co-editor and all the authors, contributors and reviewers of this specific issue that gives the reader a clear and comprehensive vision of the past and recent developments of the information and telecommunication technologies applied to the power systems.

Paris, June 2016

Philippe Adam
CIGRE Secretary General

Foreword

Since the beginning of Power System deployment, telecommunication services have been a very important component to operate the grid. Their role has evolved driven by two factors, the need to implement new grid operation capabilities, and the evolution of Information and Telecommunication technologies.

At the beginning of twentieth century, power utilities required reliable voice communications to operate the grid. Power line carrier, called in those times "wave telephony", was the first technology that contributed to the operation of the power grid.

When CIGRE was founded, in 1921, there were around 40 PLC voice links in service in USA and Europe. The deployment of this technology grew very fast and in 1929 it was reported that more than 1000 links were in operation in Europe and USA.

CIGRE identified the need for support for improving and developing communication services and a specific study committee was founded to identify issues of concern, to propose the development of new international standards and to produce guidelines on the application of new telecom and information technologies.

Many new technologies have been introduced since those early times. Study Committee D2 has actively contributed to a smooth introduction of new technologies and new ways to implement and operate these networks. These technological developments pave the way to a better control and operation of the grid, driving to an improvement of the global efficiency of the power system.

The use of fiber optics cables on overhead lines introduced a turning point in the capabilities of power utilities to develop powerful telecommunication infrastructures. Digital networks with very high transmission capacity were deployed creating a synergy between technology improvements and new operational service implementation and paving the way to the development of new operational services that offer new opportunities to implement new ways of operation and protection of the grid.

The deployment of digital networks over fiber optic infrastructure provides an almost unlimited capacity introducing the false idea that increased network bandwidth which could compensate any other issue and that a reliable operational service could be provided without any other considerations.

Study Committee D2 recognizes many other issues that should be considered when implementing reliable operational services. In the past years, several technical brochures were produced to develop these new topics. These technical brochures describe different aspects relevant to mission-critical operational service implementation including new opportunities brought about by new technologies as well as their corresponding constraints in such a way to present a complete picture of modern service provision architecture.

When CIGRE Green Books series were launched, study committee D2 identified this as a great opportunity to spread our findings and bring together several works in order to provide a comprehensive view of operational service provision in the scope of present power utilities scenario and considering the incoming Smart Grids deployments. Consequently, it was decided to initiate the production of a Green Book. This task gave us the opportunity to compile and update existing material to present it in a smooth and didactic manner.

This book is a compact study focused on describing key and specific considerations that should be taken into account to implement operational services as well as the architecture alternatives that better suit the fast evolving scenario of modern power grid and the future Smart Grid deployment. With the objective of producing a compact and focused work, technologies and international standards are not described in detail, since those details can be found in many other publications.

I am confident that, by reading the book, you will discover every relevant aspect of operational services provision in the scope of power utilities. Specific requirements and key working principles of every operational service type as well as differentiated implementation aspects are developed through the text providing a comprehensive view of different functionalities and their interactions.

This book is the result of the contribution of many specialists from Study Committee D2 and specially of Mr. Mehrdad Mesbah, Convenor of the D2 Advisory Group on Telecom, who did a great job compiling, sorting and updating information to get the text you are about to read.

It is important to point out that this is the first CIGRE Compact Studies book, a new format and idea that has seen the light thanks to the support and help of my friend Dr. Konstantin Papailiou who very cleverly saw the need and advantages of this new type of CIGRE books. I would also like to thank Mr. Mark Waldron, the Chairman of the Technical Committee of CIGRE and Mr. Philippe Adam the Secretary General of CIGRE for their support to this new format. I am sure more CIGRE Compact Studies books will follow but meanwhile enjoy this one!

Barcelona, June 2016

Carlos Samitier
CIGRE Study Committee D2 Chairman

Preface

Electrical power delivery systems are undergoing tremendous change under the impact of economic, environmental, organizational, and technological factors. This change spreads across the whole landscape of power generation, transmission, distribution, supply and consumption, bringing together these previously distinct domains into closely related processes requiring substantial exchange of information.

At the same time, energy deregulation is breaking the unique government-owned monopolistic "Vertically Integrated Utility" into separate organizations, many of which are privately controlled and operating in a competitive electricity market. This competitive environment implies a permanent quest for cost optimization through optimal usage of assets, new service provisioning processes and enhanced value proposition. It also creates the need for new coordination and control of the system operation and new regulating authorities controlling the reliability and the security of the power supply and its underlying infrastructure.

Information exchange is no longer limited to the collect of information in a unique control center and dispatch of commands to the field, but a complex inter-exchange of information between all actors of the power generation, delivery, supply, and consumption process. These actors include transmission and distribution control centers, regional and national system operators, market platforms and regulating authorities, electrical substations, bulk power plants and energy farms, distributed energy generation and storage plants, service providers, and energy consumers.

Such a complex multi-actor environment assures that the produced energy matches the consumed energy, the electrical power is transported and distributed in an optimal manner, the power system assets are used in a secure manner and maintained correctly, that energy producing entities are compensated accordingly, and that the power consumer adapts its power consumption to the realities of the generation capability at any time.

Information exchange also enables the system to adapt to any faults or anomalies in the power system in order to avoid grid instabilities and consequent power outages, assuring hence the continuity of power supply. Protection relaying of electrical transmission lines is increasingly interacting across the grid, exchanging information for higher dependability, security and selectiveness, while evolving to a more adaptive and more "grid-aware" behavior. Restoring power system assets still requires the intervention of field workers who have to detect anomalies and react rapidly to restore the system with more limited workforce and often to solve a much wider range of technical issues implying remote technical support.

A common denominator to all the mentioned changes in the power system landscape is abundant communications: between field devices, information and control platforms and utility staff, in field sites, in operational centers, or in engineering and technical support offices.

This present book is prepared with the ambition of describing the operational telecommunication networks and services of electrical power utility, the present and near-future applications for which they must assure the interconnections, their communication constraints and requirements, as well as the way they could be planned, designed, deployed and operated in a rapidly changing environment. It is based on the works of several CIGRE D2 working groups in which I have been involved in recent years and have had as a common thread the same question "how can the operational telecom network adapt to new power system communication requirements, the change of scale of the network, new communication technologies, and new organizational realities?" It is therefore the result of many contributions, discussions and experiences of a large number of utility and industry experts from many countries to whom I am particularly indebted and who are acknowledged.

It should be noted that the present book is not a textbook on telecommunications and networking technologies, for which excellent books exist already. The objective has rather been to produce in a same volume a coherent compilation of technical material, practical guidelines, process modeling and organizational principles necessary for an understanding of the communication issues in the new power utility context and orientations that have been taken or are being examined by the power system communication community worldwide.

Paris, June 2016 Mehrdad Mesbah

Acknowledgments

This book is the result of active participation of a large number of experts, listed in the appendix, who contributed through their discussions to the preparation of four CIGRE Technical Brochures, the main source of the present volume. We acknowledge their decisive role in providing the material.

We acknowledge also other CIGRE D2 Working groups in telecommunications whose work has inspired many subjects presented in this book.

We are particularly indebted to the members of CIGRE D2 Advisory Group AGD2.03 for their feedback and in particular to Mr. Hermann Spiess, a major contributor of CIGRE D2 and a long date operational telecom expert who read the draft and brought many precious suggestions and corrections to the content.

Carlos Samitier
Mehrdad Mesbah

Contents

About the Editors

Editor

Mr. Carlos Samitier holds a degree in Telecommunication and an MBA from the University of Madrid. He has been working for more than 35 years in the field of Power Utility Control Networks. He is the founder and CEO of Pullnet a company providing solutions in the field of networking and automation for Power Utilities as well as IEC 61850 tools.

He spent his professional career working in the field of Telecom for Operational Applications developing different Telecom solutions for Protection and Control applications and he has also been playing an active role in different standardisation bodies being the co-author of several standards.

He is strongly involved with CIGRE. He serves as the Chairman of Study Committee D2 on Information Systems and Telecommunication. Since 1997, he has been the member and convenor of different working groups. He has contributed actively within CIGRE publishing for more than 40 papers in international conferences and symposia. He received the Technical Committee Award in 2003 and was appointed Distinguished Member and Honorary member of CIGRE in 2012 and 2016 respectively.

Co-Editor

Mr. Mehrdad Mesbah is a senior telecom expert in charge of innovation and technical strategies in Utilities Communications for GE Energy Connections in Paris, France. He has been involved for over thirty years in telecom network design and implementation, network transformation plans, telecom architectures and technologies for numerous power utilities across the world. He has contributed to CIGRE D2 Study Committee on different aspects of operational telecoms in electrical power utilities for over a decade. In particular, he has been in charge of international working groups on operational telecom service provisioning and delivery, on utility telecom operation & maintenance and on protection relay communications. At present he leads the CIGRE D2 Advisory Group on telecommunications.

He graduated in electronic engineering from Southampton University, UK in 1982 and in telecom engineering from Ecole Nationale Superieure des Telecommunications (Telecom Paristech), Paris, France in 1984. Mehrdad Mesbah has authored many technical papers and received the CIGRE technical award for Telecommunications and Information Systems in 2008.

Abbreviations

ACL	Access Control List (Cyber-security)
ACSI	Abstract Communication Service Interface
ADSL	Asymmetric Digital Subscriber Line
ADSS	All Dielectric Self Supporting optical cable
AGC	Automatic Generation Control
AMI	Advanced Metering Infrastructure
AN	Application Network
BCP	Business Continuity Planning
BE	Best Effort (IP)
BER	Bit Error Rate
BO	Blocking Overreach Protection
BPL	Broadband Power Line Communication
BWM	Broadband Wireless Mesh
CAPEX	Capital Expenditure
CB	Circuit Breaker
CD	Collision Detection
CDMA	Code Division Multiple Access
CDP/CDR	Current Differential Protection/Relay
CIR	Committed Information Rate
CMDB	Configuration Management DataBase
COS	Class of Service
CRC	Cyclic Redundancy Check
CSMA	Carrier Sense Multiple Access
CWDM	Coarse Wavelength Division Multiplexing
DA	Distribution Automation
DCUB	Directional Comparison UnBlocking Protection
Diffserv	Differentiated services (QoS)
DM	Degraded Minutes
DMS	Distribution Management System
DMZ	De-militarized Zone
DNS	Domain Name Server
DoS	Denial of Service
DR	Demand Response
DR	Disaster Recovery

DSL	Digital Subscriber Line
DSM	Demand Side Management
DSSS	Direct Sequence Spread Spectrum
DTT	Direct Transfer Tripping
DWDM	Dense Wavelength Division Multiplexing
E1	2048 kbps G.703 interface
ECMP	Equal Cost Multipath Routing (MPLS)
EDFA	Erbium-doped Fiber Amplifier
EF	Expedited Forwarding (IP)
EFM	Ethernet in the First Mile
EHV	Extra-High Voltage
EIA	Electronic Industries Alliance (USA)
EIRP	Equivalent Isotropic Radiated Power
E-LAN	Multipoint Carrier Ethernet Service
E-Line	Point-to-point Carrier Ethernet Service
EMC	Electromagnetic Compatibility
EMS	Energy Management System
EoS	Ethernet over SDH/SONET
EPL	Ethernet Private Line
EPLAN	Ethernet Private LAN
EPON	Ethernet Passive Optical Network
EPR	Earth Potential Rise
EPU	Electrical Power Utility
ERP	Ethernet Ring Protection
ES	Errored Seconds
EVC	Electrical Vehicle Charger
EVPL	Ethernet Virtual Private Line
EVPLAN	Ethernet Virtual Private LAN
FACTS	Flexible Alternating Current Transmission System
FAN	Field Area Network
FCAPS	Fault, Configuration, Accounting, Performance, and Security
FDIR	Fault Detect, Isolate, and Service Restore
FDM	Frequency Division Multiplexing
FEC	Forward Error Correction
FEC	Forwarding Equivalence Class
FHSS	Frequency Hopping Spread Spectrum
FRR	Fast Re-route Restoration (MPLS)
FTTX	Fiber to the Home, Curb, etc.
GbE	Gigabit Ethernet
GFP	Generic Framing Protocol (Ethernet over SDH)
GIS	Geospatial Information System
GOOSE	Generic Object Oriented Substation Event
GPRS	General Packet Radio Service
GPS	Geostationary Positioning Satellite
GSE	Generic Substation Event

GSM	Global System for Mobile Communications
GSM-R	GSM for Railways
HAN	Home Area Network
HDSL	High-speed Digital Subscriber Line
HMI	Human Machine Interface
HRP	Hypothetical Reference Path (ITU-T Y.1541)
HSDPA	High-Speed Downlink Packet Access
HTTP (-S)	HyperText Transfer Protocol (-Secure)
HVAC	Heating, Ventilation, Air-Conditioning
HVDC	High Voltage Direct Current
ICCP	Inter-Control Center Protocol
ICMP	Internet Control Message Protocol
ICT	Information and Communication Technology
IDS	Intrusion Detection System
IDU	Indoor Unit
IEC	International Electrotechnical Commission
IED	Intelligent Electronic Device
IEEE	Institute of Electrical and Electronic Engineering
IETF	Internet Engineering Task Force
IGMP	Internet Group Management Protocol
IP	Internetworking Protocol
IPG	Inter-Packet Gap
IPS	Intrusion Protection System
IT	Information Technology
ITIL	Information Technology Infrastructure Library
ITU-T	International Telecommunication Union
JBD	Jitter Buffer Delay
L2TP	Layer 2 Tunneling Protocol
LAN	Local Area Network
LCAP	Link Aggregation Control Protocol
LCAS	Link Capacity Adjustment Scheme
LDP	Label Distribution Protocol (MPLS)
LMR	Land Mobile Radio
LN	Logical Node
LOS	Line-of-Sight wireless link
LSP	Label Switched Path
LSR	Label Switch Router
LTE	Long-Term Evolution Cellular Mobile Network (4G)
MAC	Medium Access Control
MAN	Metropolitan Area Network
MEF	Metro Ethernet Forum
MIB	Management Information Base
MM	Multimode Fiber
MMS	Manufacturing Message Service
MPLS	Multi-Protocol Label Switching

MPLS-TP	Multi-Protocol Label Switching - Transport Profile
MSP	Multiplex Section Protection
MSTP	Multiple Spanning Tree Protocol
NBI	North-Bound Interface (Network Management)
ND	Network Delay
NERC	North American Electric Reliability Corporation
NERC-CIP	NERC Critical Infrastructure Protection
NGOSS	Next Generation Operation Support System
NLOS	Non-Line-of-Sight wireless link
NMS	Network management system
NNI	Network to Network Interface
NOC	Network Operation Center
NTP	Network Time Protocol
O&M	Operation and Maintenance
OAM	Operation Administration and Maintenance
ODF	Optical Distribution Frame
ODU	Outdoor Unit
OFDM	Orthogonal Frequency Division Multiplexing
OMS	Outage Management System
OPEX	Operation Expenditure
OPGW	Optical Ground Wire
OSPF	Open Shortest Path First
OSS	Operation Support System
OT	Operational Technology
OTN	Optical Transport Network (ITU-T)
PD	Packetization Delay
PDC	Phasor Data Concentrator
PDH	Plesiochronous Digital Hierarchy
PDV	Packet Delay Variation
PED	Packet Encapsulation Delay
PER	Packet Error Ratio
PHB	Per Hop Behavior
PHEV	Pluggable Hybrid Electrical Vehicle
PHP	Penultimate Hop Popping (MPLS)
PING	Packet Internet Groper (ICMP echo request)
PIR	Peak Information Rate
PLC	Power Line Carrier
PLR	Packet Loss Ratio
PMR	Private Mobile Radio
PMU	Phasor Measurement Unit
POE	Power over Ethernet
PON	Passive Optical Network
POS	Packet over SONET (similar to EoS)
POTT	Permissive Overreach Transfer Tripping
PPP	Point-to-Point Protocol

PRC	Packet Route Control
PRP	Parallel Redundancy Protocol
PRR	Packet Route Restore Time
PSTN	Plain Switched Telephone Network
PTD	Packet Transfer Delay
PTP	Precision Time Protocol (IEEE 1588)
PUTT	Permissive Underreach Transfer Tripping
PW	PseudoWire
PWE	PseudoWire Emulation
PWE3	Pseudowire Emulation Edge-to-Edge
QAM	Quadrature Amplitude Modulation
QD	Queuing Delay
QoS	Quality of Service
RADIUS	Remote Authentication Dial-In User Service
RAS	Remedial Action Scheme
RCA	Root Cause Analysis (Fault Management)
RF	Radio Frequency
RFC	Request for Comments (IETF)
RMON	Remote Monitoring
RSTP	Rapid Spanning Tree Protocol
RSVP	Resource Reservation Protocol
RTU	Remote Terminal Unit
SAN	Storage Area Network
SAP	Service Access Point
SAS	Substation Automation System
SAToP	Structure-Agnostic Time Division Multiplexing over Packet
SCADA	Supervisory Control And Data Acquisition (system)
SDH	Synchronous Digital Hierarchy
SES	Severely Errored Seconds
SFD	Store-and-Forward Delay
SHDSL	Single-pair High-Speed Digital Subscriber Line
SIPS	System Integrity Protection Scheme
SLA	Service Level Agreement
SM	Single Mode Fiber
SNCP	Sub-Network Connection Protocol
SNMP	Simple Network Management Protocol
SNR	Signal-to-Noise Ratio
SNTP/NTP	(Simple) Network Timing Protocol
SOC	Security Operation Center
SONET	Synchronous Optical Network
SPS	System Protection Scheme
SSDSL	Synchronized Symmetric Digital Subscriber Line
SSH	Secure Shell
SSL	Secure Socket Layer
STP	Shielded Twisted Pair

STP	Spanning Tree Protocol
SV	Sampled Value
SyncE	Synchronous Ethernet
TACACS	Terminal Access Controller Access control System (Cyber-security)
TASE-2	Telecontrol Application Service Element
TCP	Transmission Control Protocol
TDD	Time Domain Duplexing (Ping-Pong)
TDM	Time Division Multiplexing
TDMA	Time Division Multiple Access
TE	Traffic Engineering
TETRA	Terrestrial Trunked Radio (previously Trans-European Trunked Radio)
TI	Time Interval
TLS	Transport Layer Security
TMF	Telemanagement Forum
TN	Transport Network
TP	Teleprotection Signaling
TPC	Twisted Pair Cable
TS	Time Slot
TSP	Telecom Service Provider
UA	User Application
UDP	User Datagram Protocol
UHF	Ultra-High Frequency Radio
UMTS	Universal Mobile Telecommunication System
UNI	User to Network Interface
UTC	Utilities Telecom Council
U-Telco	Utility Telecom Commercial Service Operator Company
uTOM	Utilities Telecom Operation Map
UTP	Unshielded Twisted Pair
VC	Virtual Container
VCAT	Virtual Concatenation
VDSL	Very High bitrate Digital Subscriber Line
VID	VLAN Identifier
VLAN	Virtual Local Area Network
VoIP	Voice over IP
VPLS	Virtual Private LAN Service
VPN	Virtual Private Network
VSAT	Very Small Aperture satellite Terminal
VVC/VVO	Volt-VAR Control/Optimization
WAMS	Wide Area Monitoring System
WAN	Wide Area Network
WAP&C	Wide Area Protection and Control
WDM	Wavelength Division Multiplexing
WiFi	Wireless Fidelity

WiMAX	Worldwide Interoperability for Microwave Access
WLAN	Wireless Local Area Network
WMAN	Wireless Metropolitan Area Network
WPA/WPA2	WiFi Protected Access
WPAN	Wireless Personal Area Network
WSN	Wireless Sensor Network

Introduction

Electrical Power Utilities (EPUs) are increasingly dependent upon the existence of fast, secure and reliable communications services. These services interconnect the actors, platforms, and devices constituting the technical, commercial, and corporate processes of the utility across its different industrial and corporate sites. The communication services are provisioned, managed, and maintained in different ways depending upon different quality constraints, cost and regulatory imperatives and company policy considerations. The services can be integrated together into a common network or provisioned through completely separate networks. The associated telecommunication organization of the utility varies correspondingly among power utilities.

This green book is about the specification and provisioning of communication services for the electrical power utility operational applications, the consequent deployment or transformation of networks to deliver these specific communication services, and finally the way the network and its services can be monitored, operated, and maintained to assure that the high quality of service necessary for the operation of the power system is achieved. It is composed of 5 parts as presented in Fig. 1.

Implementing telecom networks in the power utility is neither a top-down nor a bottom-up process: we cannot dimension the network from the sum of requirements and we cannot build the network and see which application requirements are fulfilled. There is a circular relationship between requirements and capabilities!

The network is planned and deployed without full and precise knowledge of all applications that it shall be serving over its useful life-time. The communication requirements of existing applications are growing with the new data transmission capabilities of the telecom network. An operational department cannot therefore give a very reliable projection of its application's exchange requirements. Similarly, "distributed intelligence" applications are designed according to data throughput, time delay and dependability levels that the network can assure.

At the same time, the power system telecom network and its potential user applications are much more intimately coupled than in the case of a commercial multi-purpose telecommunication infrastructure and must be designed to cover all requirements. As an example, power system protection relaying represents an insignificant proportion of the network's traffic but cannot be discarded in the

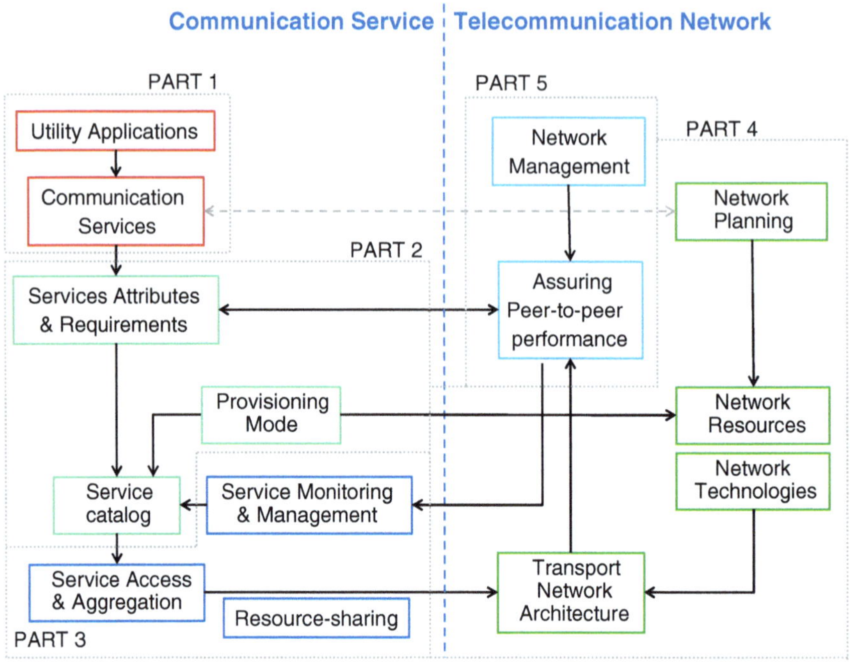

Fig. 1 Implementation process for utility telecom services and infrastructures

Electrical Utility telecom network design! By comparison, a telecom operator's service catalog covers only the requirements of an economically viable portion of the total potential demand.

Another important issue in this respect is the relative time-scales of applications deployment and network deployment. Considering the extremely rapid pace of evolution in communication technologies an early network deployment for far future applications results in unused capabilities and therefore investments and may still result in costly and yet unsuitable services when the applications are finally deployed as cost and capabilities of telecom technology evolves in the meantime. On the other hand, a network cannot be planned, designed and deployed overnight to provide the necessary services adapted to new applications, new regulatory constraints, or new organizational requirements.

The same arguments may be applied to the processes, organizational changes, and tools for the management of the telecom network and service. These need to be adapted to size, type, and scope of the telecommunication services and infrastructures. One cannot undertake major modifications in view of far future network expansions but a change of scale cannot be produced spontaneously.

As a conclusion, network planning, communication requirements assessment, and management process design are continuous and iterative processes performed concurrently as shown in Fig. 1.1. These tasks necessitate continuous examining of time-scales, technological and economical surveys, and organizational and regulatory evolutions. An appropriate network development plan must comprise steps and scenarios to allow gradual migration of a network to fulfil new requirements. The architecture must be designed in such a way to accept predictable change and capable to adapt to unpredicted accelerations or change of direction.

The present book is dedicated to operational communications as defined in the following sections. Not only the scope and mission of CIGRE is limited to these communications, but also the general-purpose enterprise network specification, design and deployment issues are abundantly covered in existing telecom networking literature.

Corporate enterprise communications covering the administrative applications of the utility organization as well as business and market communications between utility's IT platforms and external power delivery stake holders are therefore excluded from our analysis.

The integration of IT and OT (Operational Technology) data traffic over a unique network is subject of debate across the power community. The economy of scale in terms of infrastructure and management effort, sought through convergence, is to be reconciled with the requirement for cyber-security in the operational environment, difference in applications life cycles, as well as resilience to change and service outage. An assessment of opportunities and risks for the integrated transport of Enterprise and Operational networks (IT and OT networks) is given in Chap. 20.

One should however note that new applications at the frontier of Enterprise and Operational networks, Technical Office-to-Field Process applications, constitute one of the main paradigms of the new utility communications. These IT/OT applications are described and taken into account as part of the general service requirements. Their secure deployment is discussed in the respective sections of the entire book.

The book being largely the synthesis of several works performed in CIGRE D2 working groups over the recent years, very large extracts of the corresponding CIGRE Technical Brochures have been reproduced without major change, although re-organized, updated, and completed as necessary to fill detected gaps, to eliminate redundancies and finally to make it comprehensive and fluid. References used by the Technical Brochures have not been cascaded in the present volume and can be found for further reading in the original documents. I hereby acknowledge the usage of this great number of publications. The list of CIGRE Technical Brochures used as source to the book is produced in the appendix.

Part I
Operational Applications and Requirements

1 Operational Applications in the Power Delivery System

The operation of the electrical power system requires the exchange of information between different constituents of the system. The exchanged information is constantly evolving from simple signals such as commands and binary status data toward more elaborate forms such as coded measurements, control messages, and enriched situational information. Power system operation also necessitates the interaction of human operators across the system as well as their interaction with field-located power system components or remotely located operational information systems. From original control room voice communication covering power plants, substations, and load dispatch centers, the power system operation is moving to more elaborate modes of communication such as mobile data systems, remote access to field assets, and field worker access to central support platforms and information systems.

The evolution of information exchange across the power system needs to be assessed in order to determine the size, type, and perimeter of the telecommunication network that the Electrical Power Utility (EPU) should implement whether this is achieved through the deployment of a dedicated infrastructure or through provisioning of services over the network infrastructure of a telecom provider.

Figure 1.1 provides a basic model defining "user applications" interacting through a communication service delivered over a dedicated telecom infrastructure or provisioned through an operator (procured service). The Service Access Point is the point of delivery, monitoring and management for the communication service and the interface between service user and provider.

Communication services in the EPU are generally identified with the applications they serve (e.g., SCADA or protection communication services). This is indeed based on a common understanding of the telecom service provider and the application owner of the communication requirements. This is true when dealing with long-established power system applications delivered by internal telecom facilities. The same cannot be assumed when new applications are being introduced or when provisioning services from an external provider. This first part aims to

© Springer International Publishing Switzerland 2017

C. Samitier (ed.), *Utility Communication Networks and Services*,
CIGRE Green Books, DOI 10.1007/978-3-319-40283-3_1

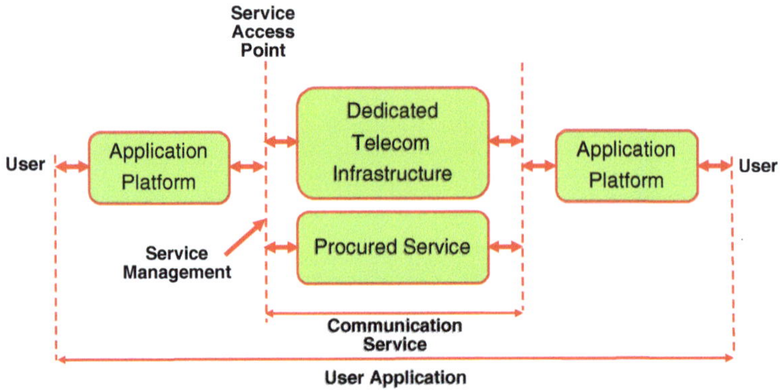

Fig. 1.1 Basic model representing the communication of user applications

characterize the main operational applications in the power system context and their respective communication service requirements.

There are no clear-cut definitions to such words as "operational" or "mission-critical." The perimeter is moving in time and varies from one utility to another. Broadly speaking, we shall limit the scope of our analysis to communication services enabling the coordination and exchange of information between the staff, devices, and platforms directly involved in applications and processes used to operate, control, and protect the power system and its constituents. These applications and processes are necessary for the proper accomplishment of the Power Utility's primary mission and therefore the associated communication services are referred to as "mission-critical." Operational services are generally required between electrical process plants (e.g., protection relay communications) or between control and monitoring platforms and the process plants (e.g., SCADA communications).

Closely related to the power system operation, there exists increasingly a group of applications related to the maintenance and support of the power system infrastructure. These operation support services include voice and data connectivity for the operation and maintenance staff, the remote monitoring and surveillance of field assets and premises to assure their working condition as well as their physical and cyber security. The operational safety and the security of the power utility industrial installations and the control of their environmental impact are subject to growing concerns and critical infrastructure regulatory constraints. Many new applications with extensive communication volume are emerging to assure these aspects. Operation Support applications often connect field-located peers (staff and device) with central monitoring and decision platforms, management and support staff, and associated information systems in utility offices.

Criticality of communication services in the Power Utility can be assessed through the consequences of service loss and degradation. It is clear that a high degree of criticality can be attributed to the operational and operation support services. However, it should be noted that these applications are not the only critical processes in the Power Utility. The financial consequences of a loss of communication in utility business and market activities can be tremendous. "Non-operational" communications are, however, more tolerant to programmed service unavailability, allowing IT-type procedures for service restoration and maintenance.

In the following sections electrical power utility operation-related applications are categorized according to their information exchange perimeters:

- **Substation-to-substation applications**—These applications comprise fast automation systems providing protection and control of power system primary assets and prompt detection and isolation of electrical faults. They assure the secure and stable operation of the electrical power system with a reaction time from a fraction of a cycle to a few cycles (power frequency cycle at 50 Hz: 20 ms). Although the term substation is used, the same principles apply to similar applications relating to bulk generation plants, energy farms, and energy-related process sites. These applications are treated in more detail as their particular communication requirements constitute one of the main drivers for implementing dedicated telecommunication infrastructures in electrical power utilities.
- **Field device to central platform applications**—These applications present status and measurement data from the grid or bulk generators to situational awareness platforms and transmit operator-initiated or application-generated commands to adjust the situation. The most typical application in this category is the Power System SCADA. Other operational applications such as Wide Area Monitoring systems (Synchronized Phasor Measurements) are increasingly deployed across the power system. Various alarm and event management platforms, supervision and monitoring of assets, site facilities, telecommunications, physical and cyber security, etc., are in this same category. Collected data is typically refreshed every 1–5 s but platform-to-field command transmission has tighter requirements.
- **Inter-platform applications**—These applications comprise the synchronization or coordination of data bases in geographically remote platforms whether for redundant processing, or in a hierarchical application (e.g., regional and national), or in a functionally distributed system (e.g., monitoring and control platforms). They generally require the transfer of much larger volumes of data but less often (and more sporadic). The time constraints here are around a few seconds. It should, however, be noted that cyber security constraints in this case become more severe in particular when more than one company and one dedicated network is concerned.
- **Office to Field applications**—These "IT-to-OT" applications cover the access requirements from an engineering office located in the corporate enterprise environment of the power utility to devices, data or staff located in the industrial process sites (field sites) of the company. It covers such applications as remote configuration, maintenance, and parameter setting of intelligent devices in the substation as well as field worker mobile data access to support platforms. The secure access from technical office to substation-based peers constitutes a major

issue of concern and a field of future development in many power utilities and the subject is hence covered in more detail.

- **Producer/Consumer to Utility Platform Applications**—A large number of new applications are planned and gradually deployed to allow the interaction of the power company with dispersed energy consumers, producers, and storage facilities. These applications comprise smart metering at energy customer's point of delivery but also the control of electrical loads and distributed power generators, public facilities such as electrical vehicle chargers and public lighting, as well as industrial and commercial/residential micro-grids. This mix of revenue management and operational information exchange is the subject of many dedicated reports and is treated here mainly with a focus on the operational side although the two information flows cannot always be fully distinguished.
- **Energy Farm Communications** are basically the same as bulk generation with more extended local communications. Some specific safety applications exist in offshore installations.

IEC 61850 Communication Model

2

Before discussing operational applications over utility telecom networks, a brief introduction is necessary on IEC 61850 standards. The initial objective of this standardization effort was to define vendor-independent data exchange architecture for the electrical substation automation system allowing interoperability of devices from different vendors. It defined substation automation data and application models and adopted standard existing communication protocol stacks and services. Over the years, IEC 61850 has grown out of the substation boundaries to cover substation-to-substation, substation-to-control center, asset condition monitoring, and synchrophasor applications, as well as communications in other power system domains. At present IEC 61850 standards are an architectural reference for data modeling of new communicating applications.

The IEC 61850 standards cover the networking of substation automation devices enabling Intelligent Electronic Devices (IEDs) to exchange information (i.e., interoperability). It comprises the following:

- a data model defining Logical Nodes (LN),
- data exchange services (Abstract Communications Service Interface or ACSI) mapped to a communication stack (MMS, TCP/IP, and Ethernet with priority tagging).

IEC 61850 is initially introduced at substation level (i.e., intra-device exchanges over a LAN). However, extensions beyond substation boundaries over a WAN have been specified and published and still ongoing. This allows, in particular, the interconnection between IEC 61850 islands.

Some of these standards and guidelines are listed below:

- IEC 61850-90-1 Use of IEC 61850 for the communication between substations
- IEC 61850-90-2 Use of IEC 61850 for the communication between control centers and substations
- IEC 61850-90-3 Using IEC 61850 for Condition Monitoring

© Springer International Publishing Switzerland 2017

C. Samitier (ed.), *Utility Communication Networks and Services*,
CIGRE Green Books, DOI 10.1007/978-3-319-40283-3_2

- IEC 61850-90-4 IEC 61850—Network Engineering Guidelines
- IEC 61850-90-5 Use of IEC 61850 to transmit synchrophasor information according to IEEE C37.118
- IEC 61850-90-6 Use of IEC 61850 for Distribution Feeder Automation System
- IEC 61850-7-410 Hydroelectric Power Plants—Communication for monitoring and control
- IEC 61850-7-420 Communications systems for Distributed Energy Resources (DER).

Figure 2.1 presents the interaction between the data model and the communications inside the IEC 61850 model. It should be noted that the standard largely leaves the implementation and architectural aspects of the communication network to the designer's initiative. A companion guideline document titled "Network Engineering Guidelines" is to provide architectural and performance analyses for both local and wide area communication networks.

The IEC 61850 standard enables information exchange through different communication services (Fig 2.2):

- Sampled Values (SV) are encapsulated and transmitted as a multicast service over Ethernet providing fast and cyclic exchange of voltage and current measurement values for protection and control replacing traditional analog wiring. Any sample loss or a delay longer than 4 ms between two consecutive samples prevents IEDs from functioning correctly. For example the Busbar voltage used to trigger protection relays is measured at 4000 samples/s and transmitted cyclically at 1 kHz. MAC-layer multicast addressing is used to make the sample values available to multiple users.

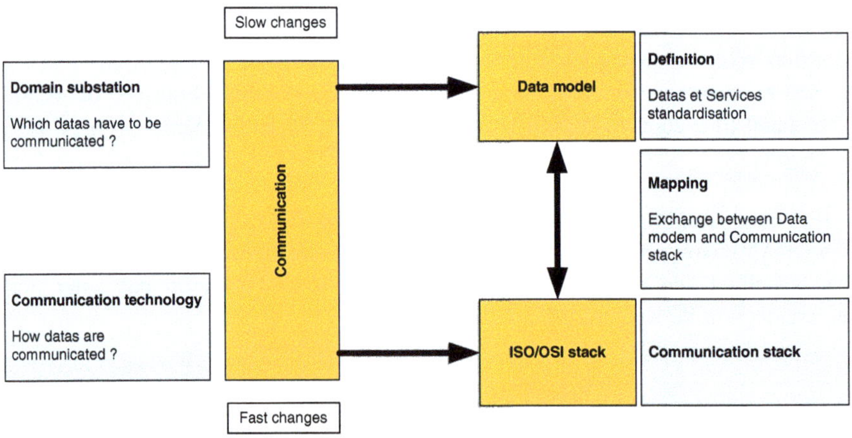

Fig. 2.1 IEC 61850 model presenting data model and communication stack

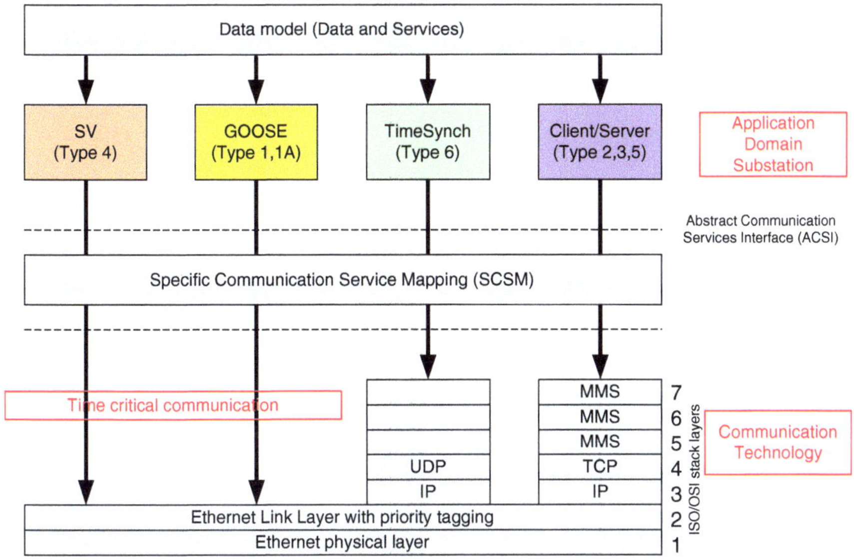

Fig. 2.2 IEC 61850 message services

- GOOSE (Generic Object Oriented Substation Event) is used for transmission of critical events in real time (e.g., tripping commands) between two or more IEDs using Ethernet multicast. The GSE service model of 61850-7-2 details fast and reliable system-wide distribution of input and output data values using a specific scheme of retransmission to achieve the appropriate level of reliability.
- TCP/IP-based messages using MMS (Manufacturing Messaging Service) data presentation layer are employed for download and upload of configuration, parameter setting, monitoring, etc.

Substation-to-Substation Applications 3

3.1 Line Protection Applications

Power system faults disrupt normal electrical power flow by diverting current through a short-circuited connection and collapsing power system voltage. Prompt and reliable clearing of power system faults therefore is of great importance in the design and operation of the power system.

- Removing faults as quickly as possible helps safeguard personnel in proximity to installations.
- Faults can cause severe operational disturbances resulting in collapse of power delivery and blackout for regions, and, in severe cases, even for several countries. Heavy reliance of modern society on electric power consuming devices for business activities, safety, lighting, heating, communication and many other conveniences make severe disturbances and blackouts unacceptable.
- Faults can cause damage and breakdown to power apparatus such as circuit breakers, transformers, and cables. The repair work or full replacement in case of destruction is very costly and may take considerable time.
- Transients due to faults in the power system can also adversely affect sources of generation and customer loads.
- Faults may have legal and financial consequences on manufacturers, held responsible for the consequences of device nonoperation (e.g., a breaker not acting correctly), and on power companies who may be penalized by the Regulation Authority or may have to compensate customers' nonsupply of power.

Electric power system generators, transformers, Busbars, and power transmission lines are monitored by Protective Relays which are designed to detect faults and consequently, to operate isolating devices in order to interrupt damaging fault current.

© Springer International Publishing Switzerland 2017

C. Samitier (ed.), *Utility Communication Networks and Services*,
CIGRE Green Books, DOI 10.1007/978-3-319-40283-3_3

The implementation of a Protection scheme must result in dependable operation of only those relays protecting the faulted unit, and at the same time must secure nonoperation of the relays during nonfault conditions and when faults occur on adjacent power system units. This balance is met only through proper protection scheme design, proper relay and equipment selection, and proper connection and setting of these relays and equipment to achieve appropriate sensitivity and coordination.

Protection performance requirements specify the balance between these conflicting goals of dependability and security.

- **Dependability** goals require maximum sensitivity and fast response time to detect and clear all faults quickly with very low probability of "failure to act" when a fault is present.
- **Security** goals require maximum selectivity and slow response time to minimize the probability of spurious operation leading to an unwanted action on a faultless circuit. Security is an issue during fault conditions as well as during normal, faultless conditions.

When protection schemes detect a fault on the equipment or line they protect, they signal (or "trip") isolating devices, called circuit breakers, to open, in order to isolate the faulty segment of the system and restore normal voltage and current flow in the power system.

When the protection scheme and circuit breakers operate properly, the fault is isolated within the required **fault-clearing time**. Protection applied on extremely high voltage systems, where fault-clearing times are most critical, typically detect faults and operate in about one to two cycles (or even less than one cycle in certain cases). Circuit breakers generally operate in one to three cycles. The combination of high-speed protection schemes and fast circuit breakers can interrupt a fault in about two cycles, although more common fault-clearing times range from three to six cycles.

Many protection applications require the real-time transfer of electrical measurements, signals, and commands between electrical substations to enhance or to enable the trip/operate decision.

A protection system must isolate the fault within a specified "fault-clearing time" of a few—typically 5 or 6 cycles (i.e., 100–120 ms at 50 Hz). This fault-clearing time includes the times for fault detection, protection decision including any signaling and isolating device operation.

Several protective relaying applications operate without any requirement for long distance communications, in particular those related to the protection of substation units (generators, busbars, transformers, etc.). Telecom services may be needed in this case, only to command a circuit breaker at a remote end if a local circuit breaker has been economized (Direct tripping) or exists but fails to interrupt fault-currents (Breaker Failure).

On the other hand, protection schemes for HV lines generally need real-time transfer of electrical measurements, signals, and command information with the protection device at the far end of the line to enhance or to enable the protection decision to operate and hence to meet fault-clearing requirements. In this case, communication may be the basis for fault detection (e.g. Current Differential Protection), or it may enhance time response and selectivity (e.g. Permissive Distance Protections), or still it may allow to command a remote isolating device (Circuit Breaker) for example when a local device fails to operate.

Teleprotection Signaling

If it were possible to set relays to see all faults on their protected line section and to ignore faults outside of their protected line section, then there would be no need for communication schemes to assist the relays. However, some protection relays, typically distance relays, cannot be set to "see" only the faults within a precise electrical distance from their line terminal. They are imprecise because of many factors including voltage and current transformer errors, relay operating tolerance, line impedance measurement errors and calculation tolerance, and source impedance variations. The primary relay elements used to detect line faults are therefore set to see or reach either short of the remote line terminal (this is called under-reaching), or to see or reach past the remote line terminal (this is called over-reaching).

The term "Teleprotection" refers to the communication interface of the Protection system (initially it applied to any protection scheme using telecommunications, now called telecom-assisted protection). Teleprotection signaling transforms the state information transmitted by the Protection Relay into a signal suitable for transmission over a telecommunication channel and restitution to the remote Protection Relay or remote Circuit Breaker in a secure and prompt manner. It may be integrated into the protective device, into the telecommunication access equipment, or more generally, it may constitute a stand-alone device.

This often supplements locally available data, confirming that a fault detected by at least one end, is in fact internal to the line, such that otherwise time-delayed operation may be accelerated.

In accordance with previously defined protection performance parameters, the operational performance of a teleprotection signaling system can be defined through the following parameters:

- **Security** is the ability to prevent communication service anomalies from restituting a Command at the remote end when no command has been issued. Security is expressed as the Probability P_{uc} of "unwanted commands" (command condition set at the receiving end for a duration longer than a specified limit). Security is related to the communication service integrity (error performance) and the Teleprotection Signaling system's error detection capability.
- **Transmission time** is the maximum time (T_{ac}) for the delivery of the command at the remote end, after which it is considered as having failed to be delivered. This is a constraint to the time performance of the communication service, not only in terms of nominal value but as a guaranteed limit.
- **Dependability** is the ability to deliver all issued commands at all times. It is expressed as the Probability P_{mc} of "missing commands" (issued commands not

arriving to the remote device, arriving too late or with a duration shorter than a specified limit). This sets a very severe constraint on the availability and error performance of the communication service, challenging such telecom service concepts as "errored seconds" and "degraded minutes" being counted in the available time of a communication service.

3.1.1 State Comparison Protection Schemes

State comparison protection schemes use communication channels to share logical status information between protective relay schemes located at each end of a transmission line. This shared information permits high-speed tripping for faults occurring on 100 % of the protected line.

The logical status information shared between the relay terminals typically relates to the direction of the fault, so the information content is very basic and generally translates into a "command", requiring very little communication bandwidth. Additional information such as "transfer tripping" of a remote breaker (to isolate a failed breaker) and recloser blocking may also be sent to provide additional control.

These schemes are fundamentally based on comparing the direction to the fault at one terminal with the direction to the fault at the other terminal permits each relay scheme to determine if the fault is within the protected line section, requiring the scheme to trip, or external to the protected line section, requiring the scheme to block tripping.

Even if the communication requirements for state comparison protection schemes are considerably less stringent than for Analogue Comparison Protection schemes (described in the next section), the "command transmission time" is of great importance because the purpose for using communication is to improve the tripping speed of the scheme. Also, variations in transmission time are better tolerated in state comparison schemes than in the Analogue Comparison protection schemes.

Communication channel security is essential to avoid false signals that could cause incorrect tripping, and communication channel dependability is important to ensure that the proper signals are communicated during power system faults, the most critical time during which the protection schemes must perform their tasks flawlessly.

Communication for state comparison protection schemes must therefore be designed to provide safe, reliable, secure, and fast information transfer from one relay scheme to another. The communication scheme must—for the vast majority of protection schemes—also be able to transmit information in both directions at the same time. The amount of information required to transfer between relay schemes depends on the relay scheme logic.

The terminology used to describe these state comparison protection schemes is basically defined according to the *impedance zone monitored by the protection relay* as presented below:

- Directional comparison blocking schemes (also called Blocking Over-reach, BO)
- Directional comparison unblocking schemes (DCUB)
- Permissive over-reaching transfer trip schemes (POTT)
- Permissive under-reaching transfer trip schemes (PUTT)
- Direct transfer tripping (DTT).

Directional Directional Comparison Blocking (BO)

In a Blocking scheme, a fault detected by a time-delayed over-reaching relay (set directional into the line) is assumed to be internal to the line unless *blocked by the device at the opposite end* (Fig. 3.1). The blocking signal indicates that a fault external to the protected line has been detected. As the blocking command is used to prevent tripping, it is critical that the communication channel should be fast and dependable. However, when a blocking signal is transmitted the line is healthy and presents normal noise and attenuation conditions to a line-correlated communication channel such as power line carrier. Blocking schemes require only a simplex signaling channel and in this case they can be applied to a multiterminal line. If the Blocking signal fails to be received during a fault, tripping will still occur for faults along the whole of the protected line, but also for some faults within the adjacent line sections.

Directional Comparison Unblocking (DCUB)

This scheme was originally applied to frequency shift keyed PLC in circumstances where faults on the carrier phase can cause significant attenuation of the PLC signal. The scheme is basically a POTT scheme supplemented with monitoring of the PLC guard channel, such that a fault detection by the over-reaching element, coincident with the absence of either a receive signal or of the guard tone; is a sufficient condition to consider the fault to be internal to the line, permitting accelerated tripping (Fig. 3.2).

Permissive Over-reach Transfer Tripping (POTT)

A time-delayed forward directional (into the line) over-reaching relay is set at each end of the line. Over-reaching means detecting line faults on a distance longer than the protected line section. When the devices at all ends of the line see the fault as forward, then it must be internal to the line and accelerated tripping of the over-reaching element may take place. A phase-segregated POTT scheme provides additional security where single pole tripping is required (Fig. 3.3).

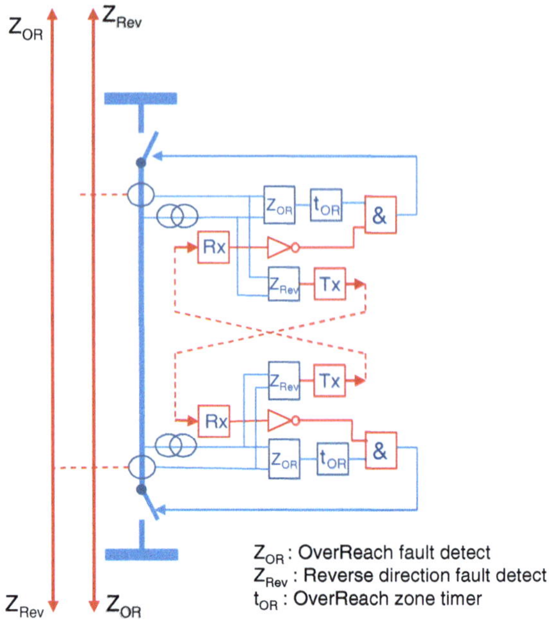

Fig. 3.1 Directional comparison blocking scheme (BO)

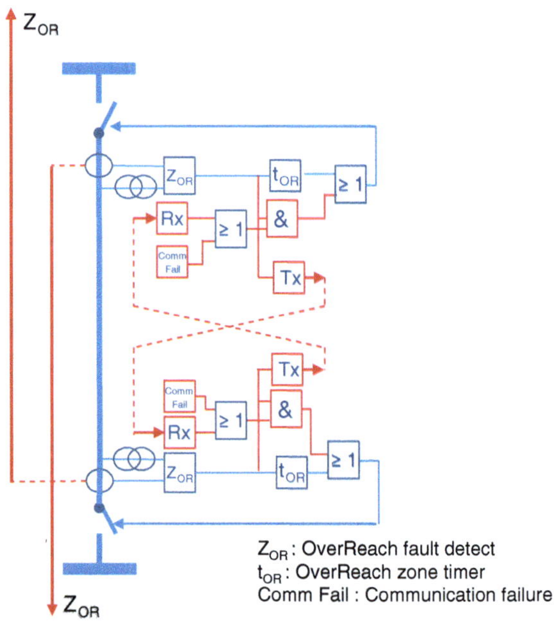

Fig. 3.2 Directional comparison unblocking (DCUB)

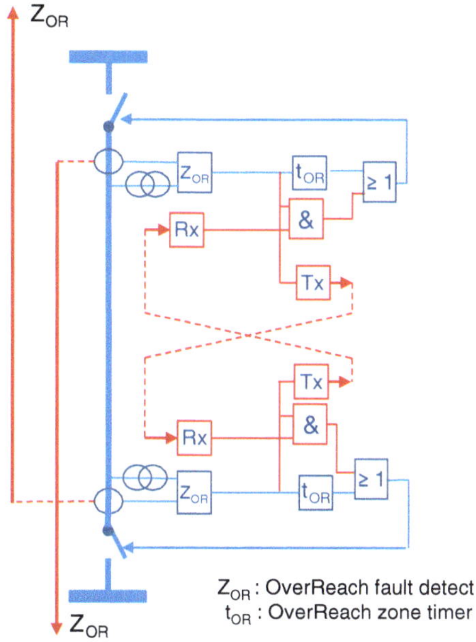

Fig. 3.3 Permissive over-reach transfer tripping (POTT)

A Permissive Over-reach (POTT) relaying scheme requires a duplex signaling channel to prevent possible false operation due to spurious keying of the signaling equipment. This scheme may be more appropriate than a Permissive Under-reach scheme (PUTT) for the protection of short transmission lines due to greater resistive coverage. It should be noted that basic distance scheme tripping will still be available in the event of signaling channel failure.

Permissive Under-reach Transfer Tripping (PUTT)

In a Permissive Under-reaching scheme (Fig. 3.4), relays at the line terminations are set to detect faults over a distance shorter than the protected line section but longer than the middle of the line section A fault detected within a non-delayed under-reaching distance protection zone is definitively internal to the line: the Circuit Breaker local to the protection, may be tripped without delay. PUTT uses fault detection of this under-reaching zone to permit tripping of Circuit Breakers at the remote ends. If the protection at any of the remote ends receives the permissive transfer tripping command and has also detected the fault, but within a time-delayed over-reaching zone then accelerated tripping is permitted at that end (Zone Acceleration). The signaling channel is only keyed for faults within the protected line giving the scheme a high degree of security. If the signaling channel fails, basic distance scheme tripping will still be available.

Assuming that the Zone 1 is set to 80 % of the protected line, the faults in the remote 20 % of the line will be cleared via the Zone 2 time delay of the local relay if the remote terminal of the line is open.

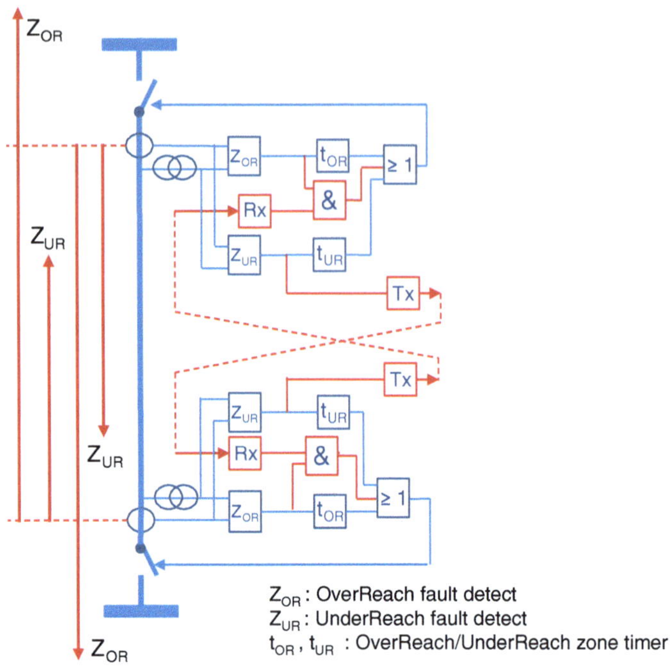

Fig. 3.4 Permissive under-reach transfer tripping

Direct Transfer Tripping (DTT)

In direct transfer tripping (or inter-tripping) applications, the command received from the far end is not supervised at the receiving end by any protection relay and simply causes Circuit Breaker operation. Since no local condition is available for checking the received command signal, it is absolutely essential that no stray unwanted commands are detected due to noise and error over the signaling channel. In other words, an inter-tripping channel must be very secure.

3.1.2 Analogue Comparison Protection Schemes

Analogue Comparison Protection is based on the transmission and comparison of electrical parameters between the ends of a protected line. The "analogue" values that are compared across the line are most commonly current samples, although other schemes such as pilot-wire or phase comparison have also been used (mostly in the past).

3.1.2.1 Current Differential Protection Schemes

The current differential protection, extensively used on HV transmission lines, compares the amplitude and phase of the local terminal currents with the amplitude and phase of the currents received from the remote terminal through a communications channel. It is applicable to any overhead line or underground cable at all voltage levels and is used in particular for the following:

- Very short lines and cables where the low impedance makes the adjustment of settings difficult for the use of Distance Relay.
- Multiterminal lines where the intermediate in-feeds modify the impedance seen by the Distance Relays, implicating that the observed impedance is not only dependent on the distance to the fault, but also on the in-feed from the remote terminals, making impossible an accurate measure of the impedance.
- HV Lines with only current transformers installed at each end (no voltage transformers).
- EHV transmission lines where series capacitors may create protection issues.
- Situations where immunity to power swings and current reversal is needed.

The currents at each line terminal are sampled, quantified, and transmitted to the remote end of the protected line for comparison. Current samples collected from a remote point must be compared with those measured locally at the same instant of time (synchronized sampling). Alternatively, the phase angles of the computed phasors are adjusted for the sampling skew and communication channel delays before comparison (asynchronous sampling). An error in sample timing and the associated delay compensation mechanism, results in a differential current that increases the risk of unwanted tripping.

Delay compensation in differential protection relies on the existence of a common time reference. Originally, and still in the great majority of installed and operational devices, this is achieved using a "ping-pong" technique to evaluate the "round-trip" transfer time and perform delay compensation assuming equal send and receive path delays. This creates a great sensitivity of the system to any time difference and therefore implicates the same routing for the two senses of communication. Moreover, considering the frequency of occurrence for this delay estimation, the communication medium must have a fairly stable propagation delay.

Modern protection systems can receive an external time reference such as a GPS clock to provide global time-stamping of samples, enabling them to tolerate switched networks. Once the system is synchronized, loss of GPS can be tolerated on a switched communication network using various techniques provided that the communication path does not change too often. There is, however, some reticence to render the protection system dependent upon an "externally provided" synchronization service such as GPS satellite.

New generation Current Differential Relays using IEC 61850 network interface and Precision Time Protocol (IEEE 1588v2) are expected to receive self-provisioned time coherence through the network which will perform the necessary ping-pong from switch to switch to determine the travel time from relay to relay. It is expected that precision time distribution will become by itself a critical “substation-to-substation” service in the future utility telecommunication network.

At present different fixed path and non-queued communications are used for Current Differential Relays. The instantaneous current sample values are converted to digital data and transmitted toward the other terminals at a rate of 12–60 samples per cycle over a communications channel of 64 kbps. Direct fiber and multiplexed communications are frequently employed. Packet-switched Ethernet communications with deterministic routing and time delay are presently being introduced. It must be emphasized that the availability of the current differential relay depends upon the availability of the communication channel. In general, the relay settings for current differential schemes are few and easy to compute, however, cable/long transmission line-charging currents and shunt-reactor applications in cables or overhead transmission circuits must be carefully studied.

There is always a finite time for the information from one end of the differential scheme to receive and process the information from the remote end and this will impact on the ability of the protection to trip. The longer the communications path and the delays in that path, the slower the overall trip time and this will become more critical as voltage levels increase. Typically, on EHV transmission systems total fault clearance times in the region of 3 cycles or less are required and given CB technology this would require unit operation in 1–1.5 cycles. This will have a direct relation with the maximum transfer time of the channel (including multiplexer, repeaters, network routing, etc.). Using differential protection requires an overall communications path delay <6 ms to achieve required fault clearance times.

Differential protection systems must have the capability of quickly detecting any loss of communications or inadequate quality of the channel in order to disable the differential scheme and employ another means of protection such as distance and over-current back up. This will indeed impact upon the achievable total fault clearance time considering the potential absence of teleprotection signaling to accelerate the back up protection. Generally, scheme designers would design mitigation strategies to reduce this risk, or use redundant communication paths/networks.

3.1.2.2 Other Analogue Comparison Schemes

Although the most commonly used analogue comparison schemes are Current Differential Protections, there are some older schemes such as AC pilot-wire relaying and phase comparison which may still be encountered as legacy applications in certain power networks. AC pilot-wire relaying is used for protecting short lines using independent metallic pilot-wires and as such do not interfere with the deployment or refurbishment of telecommunication systems. Phase Comparison Protection schemes compare the phase angles between the local and the remote terminal line currents and therefore require a communications channel to transmit and receive the necessary information. The phase comparison scheme has been very popular in the past because it had minimal communication channel requirements allowing the usage of Power Line Carriers. On the other hand, the sensitivity of the phase comparison relaying system is much lower than current differential relaying systems. Figure 3.5 provides a summary of line protection schemes described in this chapter and their communication attributes.

Protection scheme		Main Communication Attributes
Directional Comparison Blocking (Blocking Overreach)	BO	High Dependability(on healthy line), Very low transfer time,
Permissive Overreaching Transfer Trip	POTT	High Dependability, Controlled time
Directional Comparison Unblocking	DCUB	(Special case of POTT)
Permissive Under-reaching Transfer Trip	PUTT	High Dependability, Controlled time
Direct Transfer Tripping	DTT	Very High Security High Dependability, Controlled time
Current Differential Protection	CDP	Time synchronization of samples Bandwidth for transport of samples

Fig. 3.5 Summary of line protection schemes and their communication quality requirements

3.1.3 Protection Relay Communication in the IEC 61850

The following application groups have been considered in IEC 61850-90-1:

- Protection functions such as current differential line protection, distance protection with permissive and blocking schemes, directional and phase comparison protection, transfer tripping, predictive wide area protection, and substation integrity protection schemes.
- Control functions like auto-reclosing, interlocking, cross-triggering, generator or load shedding, out-of-step detection and topology determination of HV networks.

Whereas the timing requirements under normal operating conditions can be achieved within a substation, meeting the communication requirements for substation-to-substation applications such as protection relaying needs much more attention as to the WAN's time predictability and hence guaranteed QoS.

IEC 61850 data model has been designed as a self-contained object model allowing the exchange of information without prior knowledge of the exchanging device. A client can retrieve information from, and explore logical nodes using only the elements given by the node and the knowledge of the standard data model. This abstract model is the basis for the *interoperability* which has been the goal of IEC 61850. At the same time, such a self-contained data model results in a great amount of data overhead which is not required if the exchanging parties have intimate knowledge of each other allowing them to exchange in a much more compact manner (e.g., trip signals). This has led to two architectural approaches from a communication point of view as shown in Fig. 3.6:

1. Gateway (or Proxy) approach—Corresponding gateway devices (at distinct substations) exchange specific signals and messages on behalf of IEC 61850 devices at each end. This is particularly useful where only low-speed

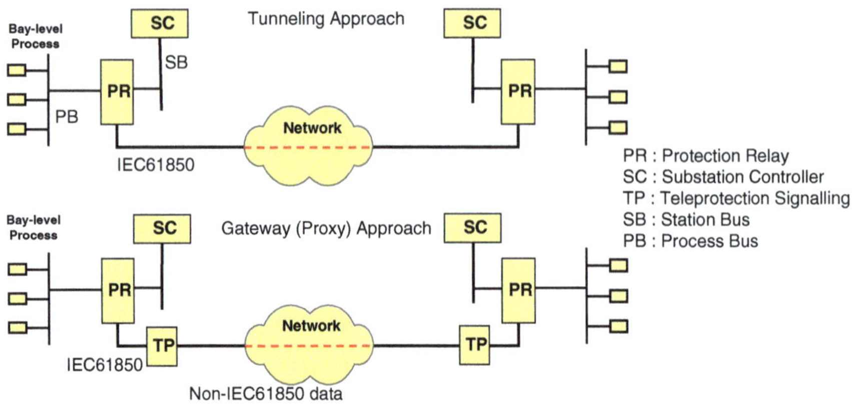

Fig. 3.6 Tunneling and gateway approaches in IEC 61850-90-1

communication channels are available. An IEC 61850-interfaced Teleprotection Signaling Equipment is typically in this category.
2. Tunneling Approach—Use high-speed communication links for direct transmission of 61850 messages from one substation to another substation.

3.2 System Protection Schemes

In many power delivery systems across the world the demand for electricity is fast approaching (and in some cases exceeding) the supply capability, while the operator still has the responsibility to provide a secure supply within statutory limits to its customers. Such increase in demand is pushing the grid to its stability limits which will ultimately result in the loss of supply or "black outs" as witnessed in recent years. Increasing the supply capability through infrastructure enhancement (e.g., new power transmission lines) has major economic and environmental implications. A mitigating approach is to implement strategies and corrective actions based on close monitoring of the variations of the power system parameters and a predetermined set of decision criteria. These systems are called System Protection Schemes (SPS), Remedial Action Schemes (RAS), or System Integrity Protection Schemes (SIPS). The following definition currently used for System Protection Schemes is given by the North American Electricity Reliability Council (NERC):

> An automatic protection system designed to detect abnormal or predetermined system conditions, and take corrective actions other than and/or in addition to the isolation of faulted components to maintain system reliability. Such action may include changes in demand, generation or system configuration to maintain system stability, acceptable voltage, or power flows.

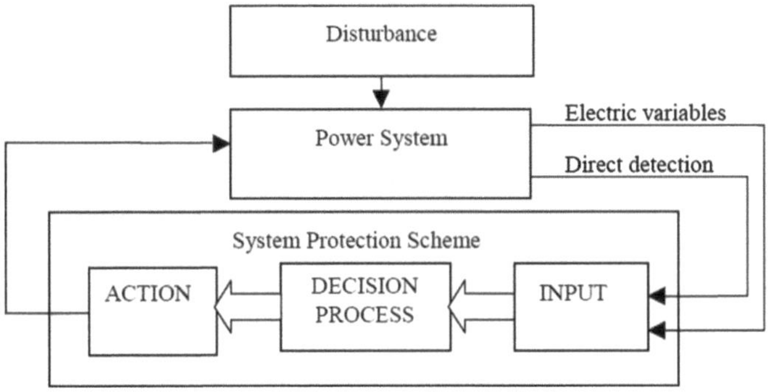

Fig. 3.7 General structure of a system protection scheme

System Protection operates over a wider portion of the grid (e.g., a transmission corridor) and a longer time (100 ms—few seconds) compared to Line and Unit Protections described previously. Typically it collects network condition information at multiple locations of a power system and undertakes system-wide automated actions such as load or generator disconnection.

Figure 3.7 shows the typical structure for a System Protection Scheme. The input to the system comprise electrical variables such as voltages, currents, and frequency measurements, but also control signals from power system stabilizers and FACTs as well as status signals from circuit breakers or tap changers. Typical actions associated to SPS decision are load shedding, generator tripping, etc.

3.2.1 SPS Applications

A large panel of power system applications responding to power system disturbance such as overload, power swing, and abnormal frequency or voltage can be implemented as System Protection Schemes. They comprise automated systems that protect the grid against system emergencies, minimizing the potential and extent of wide outages through automatic measures such as load shedding, generator shedding, or system separation. Some typical examples of System Protection applications are given hereafter together with operation time orders of magnitude in Fig. 3.8:

- Adaptive protection
- Generator control and rejection
- Load Rejection—Transmission line Removal
- Load Shedding (under-frequency or under-voltage)
- Out-of-Step Protection
- System split/separation
- Actions on Automatic Generation Control (AGC)
- VAR compensation

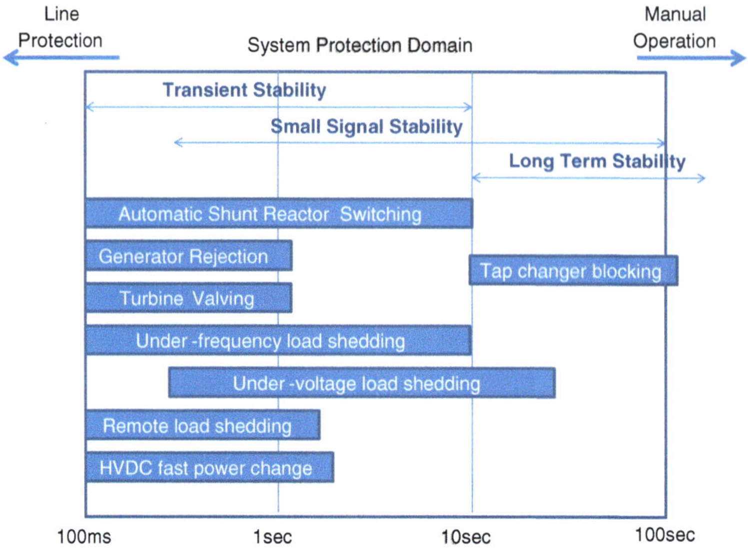

Fig. 3.8 SPS operation time frame related to power system phenomena

- Tap changer blocking
- Turbine fast valving/generator run-back
- Discrete excitation controls
- HVDC fast power change (Fig. 3.8).

3.2.2 SPS Architecture

The architecture of a System Protection Scheme can take different forms depending on the application:

- The system can be implemented with information capture, decision and action all located in electrical substations (reason why the section is inserted under substation-to-substation applications). In this case, a logic controller in one substation initiates actions across a subsystem. Reliability at the decision substation is a major design issue in this case. The decision and corrective action need secure reliable and prompt communication channels to collect remote information and to initiate actions.
- For more system-wide applications, the decision process can be totally or partially performed at a Control Center, and can initiate multilevel corrective actions. These systems are generally less time-sensitive automation applications covering stability issues in larger networks over one or multiple grids, and comprising multiple balancing authority areas. These systems are described under Substation-to-Control Center applications. The system can be composed

of a 2-level hierarchy: a lower level fast automation subsystem connecting substation sensing and actuating devices to a substation-based logic controller, and a higher level system-wide automation composed of a central platform and multiple subnetwork logic controllers.

- A System Protection Scheme can also be composed of sensing, decision, and actuating functions in the same substation together with a central coordinating platform which will adjust parameters and settings at each substation according to a wider set of situational information. An adaptive protection scheme is such a system, automatically making adjustments to protection relay settings in specific network conditions, typically to avoid cascaded tripping.

From an operational control point of view, two categories of SPS can be distinguished according to their respective control variables:

- **Response-based SPS** is based on measured electrical variables (e.g., a synchronized set of voltage phasors, frequency, etc.) and initiates actions when these variables are outside certain thresholds. Typical response-based SPS are under-frequency or under-voltage load shedding. Response-based schemes require fast exchange of large volumes of information. They require therefore extensive network communication capacity.
- **Event-based SPS** operates on detecting a combination of events (e.g., loss of several lines). Event-based SPS are faster than the response-based ones as they do not need to wait for the reception of out-of-tolerance measured variables at the decision node. Typical event-based SPS are generation rejection initiated by tripping across a transmission corridor.

In all cases, implementing System Protection Schemes requires reliable, fast, time-constrained, and fully predictable communication network services. Response-based SPS, in the form of synchrophasor-based Wide Area Protection and Control requires fast collection of a coherent data set, as described in the following paragraph, to implement complex applications. Event-based SPS, implemented using teleprotection signaling equipment and simple combinatory logics is used as a way to implement protection systems responding to simpler but higher time constraint applications.

3.2.3 Wide Area Protection & Control (WAP&C)

Response-based system protection schemes capture electrical variables across a wide geographical area and from many different sites. Analytical algorithms use the collected data set in order to adopt an automated decision which can be more or less complex depending on the reactive or proactive nature of the application. The data set is generally composed of voltage and current values across the protected system. It must have high resolution (high sampling rate) to reflect the variation waveform information (vector measurements rather than scalar amplitude values). It must also be complete (no lost values) and coherent (a snapshot of all values taken at the same instant of time). Such a data set allows performing full-scale circuit analysis well beyond the State Estimation performed by the SCADA system. Complex transform-based predictive analysis can further allow proactive measures to maintain system stability.

These System Protection Schemes are known as Wide Area Protection & Control (WAP&C). This is an extension of the concept of Wide Area Monitoring Systems (WAMS), described in the next section, for implementing Closed Loop Applications automatically acting upon the grid in a time scale longer than Line and Unit Protection but shorter than SCADA (System Protection time scale).

WAP&C systems employ time-stamped voltage and current phasors (synchrophasors) with precise time synchronization across the protected domain. This allows accurate data comparison over a wide area or across the entire controlled system. Phasor Measurement Units (PMU) are data capture devices that in conjunction with GPS provide time-synchronized vectors. The standard IEEE C37.118 defines the coding of phasor information and the associated time-stamping. Phasor Data Concentrators (PDCs) receive phasor data from the PMUs and run decision algorithms for taking appropriate actions. They can also act as gateways towards higher level information processing at a Wide Area Monitoring and Control System.

Factor	Reporting Rate range	End-to-end Latency	Measurement Timing Error	Sensitivity to Message Transfer Delay Variations	Sensitivity to Lost Packets	Currently covered in 61850
Sync-check	≥ 4/s	100 ms	50 μs	Medium	High	SV service
Adaptive Relaying	≥ 10/s	50 ms	50 μs	Low	Medium	SV service
Out-of-step Protection	≥ 10/s	50 – 500 ms	50 μs	Medium	Medium	SV service
Situational Awareness	1/s to 50/s	5 s	50 μs	Low to medium	Low to medium	Periodic reporting, SV service
State-Estimation & Security Assessment	1/300s to 10/s	5 s	50 μs	Low	Medium	Periodic reporting, SV service
Data Archiving	Any	N/A	50 μs	Low	Medium	All as needed
Wide Area Controls	≥ 10/s	50 – 500 ms	50 μs	Medium	High	SV service

Fig. 3.9 Wide area protection and control applications

A high resolution capture and tight time imperatives (for prompt closed loop operation) imply high bandwidth and low latency deterministic communications. The completeness of data sets is to be assured through the reliability and integrity of the communication system. This cannot be fulfilled by an acknowledged exchange and retransmission upon error detection due to time imperatives of the closed loop system. Deploying Ethernet transport with controlled quality of service in the communication networks allows implementing many complex system protection schemes. The power system communication architecture standard IEC 61850 provides appropriate protocol stack and services for the information exchange in the WAP&C.

The time stamp precision can be achieved either through network-independent time distribution (e.g., GPS) or through a precise clock distribution service over a packet-switched network.

IEC 61850-90-5 describes how synchrophasors can be transmitted via IEC 61850. Sample transmission is based on the Sample Value (SV) service. For a communication outside the substation this service has to be tunneled across a broadband communication network. Additional event data can be communicated via GOOSE or MMS reports.

IEC 61850-90-5 describes the communication requirements for many protection schemes based on synchrophasors. Fig. 3.9 summarizes these requirements.

4 Field Device to Central Platform Applications

This section covers a very large number of utility applications ranging from power system Supervisory Control and Data Acquisition (SCADA) to all kinds of monitoring and surveillance applications collecting data from electrical substations or other field industrial process installations for processing at some central platform and providing "situational awareness" to grid operators, grid automation applications, asset managers or security surveillance staff. The central platform can react to the outcome through transmitting command signals to modify the situation (e.g., initiating circuit breaker operations, switch-over actions, etc.), through blocking actions (e.g., cyber security monitoring), or through initiating field interventions, depending on the nature of the application. Some common denominators of these data collecting, monitoring, and supervision applications in terms of communications are as follows:

- Preestablished peers—similar to substation-to-substation applications, the communication peers are predefined and invariable—from each substation to the corresponding platform.
- Communication can be initiated by the central platform (cyclic data collection or exceptional data request) or by "satellite" substations (event-driven exchanges).
- Continuous data flow—according to an application-dependent periodicity, data is collected from each substation on a continuous and permanent basis. The data traffic in the opposite direction is much smaller (i.e., highly unbalanced). This can be for example, sending commands or initiating data requests from a platform to a field installed device.
- Constant throughput—The required network throughput for these applications is almost constant depending on the volume of database to be cyclically refreshed.
- Time sensitivity—These applications are considered as real-time, but with no critical time constraints in the data collection (uplink). When data collection is performed through a request-response polling protocol, however, the transfer time may govern the data exchange efficiency. The down-link direction (e.g., command) however, is time sensitive and may need higher time control.

© Springer International Publishing Switzerland 2017

C. Samitier (ed.), *Utility Communication Networks and Services*,
CIGRE Green Books, DOI 10.1007/978-3-319-40283-3_4

Typical examples of substation-to-platform communications are given below:

- Power System SCADA
- Synchrophasor-Based Wide Area Monitoring Systems (WAMS)
- Video-surveillance
- Power asset monitoring
- Telecom network fault and performance monitoring
- Grid-level metering
- Environmental monitoring
- Asset cyber security monitoring
- Site access control.

4.1 Power System SCADA

The power grid's SCADA communications consist in the periodic exchange of short data messages between a central platform in the Control Center and Remote Terminal Units (RTU) in electrical substations, renewable energy storage and generation plants, or any other distributed component of the power system requiring supervision and control. The messages comprise status indications, measurements, commands, set-points, and synchronizing signals that must be transmitted in real-time and requiring data integrity, accuracy, and short transfer time.

Power transmission and distribution networks SCADA generally differ in their requirements, cost objectives, and hence suitable communication solutions. The number of outstations and their corresponding size, cost, volume of traffic, and geographical dispersion are very different in the national transmission grid and in regional distribution networks. The time constraints and the required level of availability, fault tolerance, and data integrity are also different. As a consequence, transmission grid SCADA communication is often implemented through a broadband private network (e.g., optical fiber), while in distribution networks (in particular for MV grids) low capacity UHF Multiple Address Radio systems (MARS), license-free or procured wireless services (GPRS, LTE, etc.) prevail.

Still today, the widest employed communication mode for the substation RTU remains the asynchronous serial link through an RS232 interface, polled by the central control platform. The communication protocol associated to this mode has been standardized as IEC 60870-5-101 (IEC101), although many other protocols are still in use in legacy systems. The major drawback to serial communication for SCADA is indeed its lack of flexibility (e.g., for back up control center connection) and cumbersome installation in particular at the control center. Packet switching, introduced in SCADA systems since the late 1980s for more efficient usage of leased aggregated bandwidth (X25, Frame Relay, ATM, etc.) have only gained worldwide popularity with the advent of IP communications. The TCP/IP protocol IEC 60870-5-104 (referred to as IEC104) is often the migration target for SCADA systems. The use of TCP/IP enhances considerably the flexibility of the SCADA

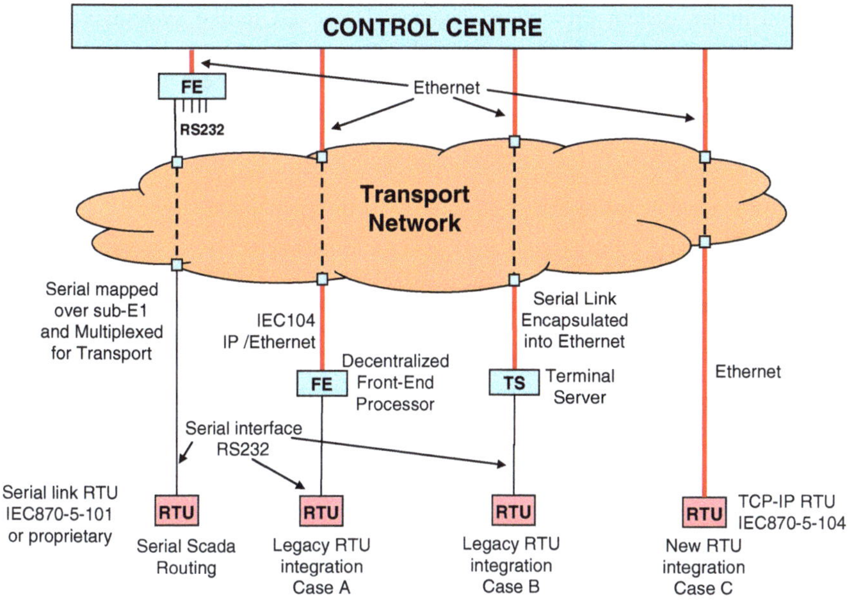

Fig. 4.1 SCADA RTU to control platform connection alternatives

communication system, facilitating the relocation of an RTU or the switch-over of RTU communications to a back up facility.

The RTU communicates through a Fast Ethernet LAN access interface, although the bandwidth allocated to each RTU communication remains often around 20 kbps. Legacy RTU may be connected through a Terminal Server or protocol gateway often integrated into the substation access device switch.

The migration process to TCP/IP in large SCADA networks may be extended over many years, with different timelines for the replacement of the RTUs, network interfaces and SCADA servers. During this long transition period, legacy and new RTUs have to coexist across the grid and the communications may be Ethernet/IP at the SCADA server but serial at the RTU end (Fig. 4.1).

4.2 Synchrophasor-Based Wide Area Monitoring System

WAMS refers to a group of power system applications which use a synchronized snap-shot of power system parameters to enable a better awareness of stability issues and power flows across a grid incorporating dispersed generation and multiple Utilities. These systems operate through phasor acquisition of parameters such as bus voltages, line currents, etc. in electrical substations The Phasor Measurement Unit (PMU) and the Phasor Data Concentrators (PDC) constitute the substation acquisition device and the concentrating gateway device respectively, as previously described under system protection schemes. However, unlike system protection schemes where decision algorithms generally in a substation provide action at other substations (therefore a substation-to-substation application), here the phasor data is generally employed in a control center for operator situation-awareness, for modeling and analysis or for enhancing the state estimator in an EMS/SCADA. Different levels of wide area applications have very different requirements in terms of information exchange and consequently telecommunication service.

- **Post-Incident Analysis and Static Modeling** applications are offline systems where collected data is used to analyze the cause of an event or to adjust the behavior model for a system. Data can be collected continuously, daily, or only on request. The communication service can be a TCP/IP file transfer service with no real-time constraint.
- **Visualization and Situational Awareness** applications collect data from sites and display them for human operator observation. These applications which constitute the great majority of present day systems have time requirements which are those of a human operator and must additionally present a level of sample loss unperceivable by the human operator. In terms of communication service a non-acknowledge UDP/IP is an adequate solution in this case whether through a dedicated network or a public provider.
- **Monitoring and Decision Support Systems** use collected data to produce analytical information helping operators respond to grid events and to position the grid for improved security and resilience. Stability diagrams and corresponding voltage collapse margins, as well as different monitoring applications (voltage and frequency stability, power oscillations, line temperature, etc.) are among these applications. Monitoring and decision support applications have time constraints which are similar to power system SCADA. This is achievable through UDP over a private IP network or a service provider VPN through a carefully specified SLA.
- **Closed-Loop Applications** are those which incorporate collecting of data from the grid, processing, automatic recognition of a pattern, and remedial action upon the grid. The systems are used for emergency situation control and special protection applications as described earlier in Sect. 3.2. Closed-loop synchrophasor applications are not yet widely implemented and their critical real-time nature necessitates particular attention on time control. Furthermore the decision to act automatically upon the network in real-time means that the data set (from different locations and sample stack from each point) must be complete, that is to say almost lossless.

PMU operation is specified by IEEE C37.118 which defines phasor construction using the GPS-satellite timing signal, as well as the phasor's data format. The exact data volume associated with the transmission of a data packet from a PMU varies depending on the incorporated parameters and the way each of them is coded (i.e., floating point or not, etc.) but can be assumed to be around 80–100 octets. This data volume is to be transferred across the network at a rate which is governed by the

	Wide Area Monitoring Applications	Latency	Resolution (sample/sec)	Comments
Situational Awareness	Situational Awareness Dashboard	1-5 sec	1	Assess system state (Normal, Alert, Alarm)
	Real-time Compliance Monitoring	1-5 sec	1	Angle of Separation, Display Voltage, Phase, Power swing, Line loading MW / MVAR flows
	Frequency Instability Detection /Islanding	1-5 sec	25-30	
Monitoring & Decision Support	Real-time Monitoring and Trending	1-5 sec	1	Decision support and security assessment Help operator to respond to grid events Provide time series information Stability diagrams & Collapse margins Reposition the grid for improved security Monitoring of voltage & frequency stability Display of line temperatures
	Real-time Alerts and Alarms	1-5 sec	25-30	
	State Estimation	1-2 min	25-30	
	Small-signal Stability Monitoring	Few sec	10-60	
	Voltage Stability Monitoring/Assessment	Few sec	25-30	
	Line Thermal Monitoring (Overload)	Few sec	25-30	
Analysis & Static Modeling	Pattern Recognition/ Correlation Analysis	N/A	1	Post-incident Analysis Identify system security metrics System-level and grid asset models
	Disturbance Analysis Compliance	N/A	25-30	
	Frequency Response Analysis	N/A	10	
	Model Validation	N/A	25-30	
Protection & Control	Adaptive Relaying	100ms	25-30	Emergency situation control and protection Closed loop SPS applications
	Out-of-step Protection	100ms	25-30	
	Small-signal stability Prot. & Control	100ms	25-30	
	Short-term stability control (e.g. transient stability)	100ms	25-30	
	Long-term stability control (e.g. Wide Area frequency/ voltage stability)	1-5 sec	25-30	

Fig. 4.2 Wide area applications communication service requirements [extracted from NASPI]

sampling frequency of the PMU. The sampling frequency is expressed as a number of (or a fraction of) AC cycles. It varies generally between 25 and 30 samples per second (one sample every two cycles) to 100–120 samples per second (2 samples every cycle). The required communication throughput is then somewhere in the range of 16–100 kbps although PDC links may require 100 kbps–1 Mbps or more.

Figure 4.2 provides some communication requirements for typical synchrophasor applications including some closed loop ones already discussed under System Protection Schemes.

4.3 Other IP-Based Monitoring Applications in the Substation

A long list of other field device to platform applications allow to monitor remotely the health of installed assets, the access security and environmental hazards in field process sites, cyber security of intelligent substation device, or the state of telecommunication devices and services in the electrical substation. These applications require access to an IP-based network connecting substations to central platforms. The exchanged data volume and hence the required communication channel throughput varies according to the quantity of information and the frequency of cyclic collection of measurements, threshold crossing events and device alarms. End-to-end latency is generally not a design or dimensioning issue for these monitoring applications although operator usage comfort still requires the systems to transmit transactional data (e.g., request/reply or command/react) in reasonable time. A number of these monitoring applications are briefly described.

Video Surveillance
Video monitoring of substations and other operational assets is a growing application in the grid due to concerns and regulatory obligations over the integrity and security of national critical infrastructures. Remote video monitoring of installations can be a source of substantial data traffic across the power network if used extensively and can drive telecom network upgrades in particular for distribution grids where communication capacity has often been very low.

Asset Condition Monitoring
Primary assets of the power system (circuit breaker, power transformer, etc.) generate condition monitoring data through their sensors and associated electronics. This data can be collected for maintenance requirements, and for determining duty cycle, device capability and loading ability. Asset condition monitoring enables the safe and efficient use of network components to their real end of life, at full efficiency, without disruption of service due to asset failure, environmental risks, or unnecessary preventive replacement.

Secondary assets of the electrical substation related to the measurement, protection, and control as well as the related power supplies can also be monitored over remote platforms. Moreover, these "Intelligent Electronic Devices" (IED) are configured, patched, upgraded, and in general, managed remotely implying not only network connectivity and bandwidth but also harsh cyber security constraints. These latter aspects are treated under Chap. 6 on "Office to field" applications.

Telecom Fault and Performance Monitoring
Telecommunication devices constitute a particular group of assets in the power system, often not directly part of power system but impacting its operational capability. Fault monitoring of telecom equipment, links and services, as well as the performance monitoring of the delivered communication services represent a nonnegligible communication load (e.g., SNMP exchanges over IP). This particular type of asset monitoring is further developed in Part 5.

Grid-Level Energy Metering and Monitoring

Energy metering information at the power delivery point of the transmission grid is required by the different actors of the open deregulated electricity market. This enables the settlement and reconciliation processes as well as invoicing of grid access services towards energy distributors. In addition to metering data, electrical power parameters are often monitored at the power delivery point to ensure the contractual quality of the delivered power.

Environmental Monitoring (Sites and Assets)

Power system process sites are increasingly equipped with environmental sensors for detecting abnormal temperature, fire, smoke, floods, gas, and chemicals and provide alarm information to remote monitoring platforms. The purpose of such monitoring is not only to protect substation assets and premises but also to protect the environment from industrial risks and hazards related to the substation assets (e.g., chemical pollution). Growing environmental concern and regulatory obligations in this field lead to steadily increasing deployments of such remote monitoring and early warning systems to avoid environmental impacts.

Asset Cyber Security Monitoring

The growing deployment of electronic intelligence in the electrical substation and the subsequent exchange of information between networked devices across the power grid have made it essential to implement numerous security barriers and intrusion detection/prevention systems. In particular, intelligent devices are made accessible from remote locations for data reporting (e.g., asset monitoring) and for remote diagnostics, parameter setting, and configuration change.

Remote access to substation device is increasingly authenticated and authorized at a remote server (e.g., RADIUS server) hence requiring particularly reliable and secure communications.

In addition to security protection at device level, remote access to the substation and to its critical information zone is further protected through white-list filtering determining the network users entitled to cross the barrier. Appropriate firewalls are integrated into the telecom access and aggregation devices and into substation switches. These security filtering components, although local, need to be remotely updated, patched and configured through reliable and secure communications. Moreover, as the number of security-related devices and systems increases, it becomes necessary to reinforce them through coordinated administration, supervision, and efficient logs processing. Security Operational Centers (SOC) are set up to collect security-related events across the network, to produce security reports and to enable the operator to take appropriate preventive measures in reaction to threat situations (e.g., blocking access ports). Security logs and reports may be used for investigations and for understanding of incidents or trends especially through dashboards.

Site Access Control

Electronic site access control systems are increasingly used to control, register, and monitor the physical access to operational sites. Smart electronic identity cards and biometric authentication are becoming part of the security and safety policy. Electronic access control allows differentiated accessibility in time and across locations for different classes of staff (operational, service contractors, maintenance, etc.).

Protecting sites like power plants or substations from intruders has always been a main concern for Utilities, not only for the site protection itself but also for human and animal safety. Fences and guards were in the past the only solutions, but with increasingly unmanned installations, intrusion detection systems are being introduced.

The classical intrusion detection system is composed of sensors (radio, laser, dry contacts, …) connected to a local collector unit that monitors the sensor states and, in case of a detection, sends online notifications to the SOC of the utility and local security forces. More recent developments use video-surveillance cameras and image analysis software that alerts SOC operators in case of image pattern changes.

Site access and intrusion detection applications require fast and reliable data communications for authentication and access registration.

5 Inter-platform Applications

Communications between Control Centers is necessary for connection to back up facilities (e.g. for database synchronization), to other Control Centers (e.g., for coordination), or to other platforms inside the perimeter of the utility (e.g., outage management) or outside this perimeter (e.g. energy market platform). The primary purpose of the Inter-Control Center Communications is to transfer data between platforms for data synchronization, status reporting, and cross-platform data access.

These communications are generally assured through the Inter-Control Center Protocol (ICCP) standardized as IEC 60870-6 (Telecontrol Application Service Element TASE-2) protocol, although earlier protocols may still be in use in certain older systems.

ICCP uses an underlying transport-service, normally TCP/IP over Ethernet. The traffic is by nature sporadic with an exchange volume which can be very large for example for main to back up control center links. The delay requirement is often of the order of hundreds of milliseconds or more depending on the applications. Lower delays may be needed for much shorter control signals. The required throughput and the consequent capacity are determined by the maximum time required for completing a large data exchange (for example a main-back up synchronization). Dedicated resources such as Ethernet-over SDH with NxE1 capacity or an Ethernet pseudo-wire over MPLS are generally allocated to these higher traffic links, although much lower capacity links have been used in the past.

Many inter-platform applications operate beyond the reach of the power utility dedicated telecom infrastructure. The communication links are therefore provisioned through public telecom service providers. Security in this case, is a more critical issue than transfer time, considering that an inadequately protected ICCP connection forms an open door to the control of the energy network.

Remote connection of Operator positions to the Control Center platform also requires high-speed communications. An Ethernet connection with a throughput of 2–10 Mbps generally allows an adequate quality communication link for connecting these remote workstations.

© Springer International Publishing Switzerland 2017

C. Samitier (ed.), *Utility Communication Networks and Services*,
CIGRE Green Books, DOI 10.1007/978-3-319-40283-3_5

6 Office-to-Field Applications

New applications are growing substantially at the interface between the enterprise and the operational worlds requiring information exchange between field operational sites and utility offices or enterprise information platforms. Some typical cases are listed below:

- Consumer smart metering applications and demand response,
- Management and control of third party renewable generation and storage,
- Market settlement and reconciliation metering at the HV substation access to the distribution,
- Asset monitoring and maintenance—Remote access from the office to grid data
- Field maintenance support—Field worker access to remote management and support platforms.

The common denominator of all these applications, which are mostly treated elsewhere in this introductory section on operational applications and their communication requirements, is that one peer of the communication is located outside the operational perimeter of electrical substations network. Figure 6.1 is a rough map of devices, platforms, and utility staff across two communication environments:

- Field Network—This is a closed network interconnecting intelligent field assets over local and wide area networks often through a dedicated communication infrastructure. It corresponds to the Operational Information Technology (OT) infrastructure and activities.
- Corporate Enterprise Network—This is a more open network for enterprise processes through which utility employees interact together, or across appropriate security barriers with peers beyond the enterprise perimeter (e.g., other utilities, system operators, suppliers, and contractors). It corresponds to the enterprise IT infrastructure and activities across the power utility.

© Springer International Publishing Switzerland 2017
C. Samitier (ed.), *Utility Communication Networks and Services*,
CIGRE Green Books, DOI 10.1007/978-3-319-40283-3_6

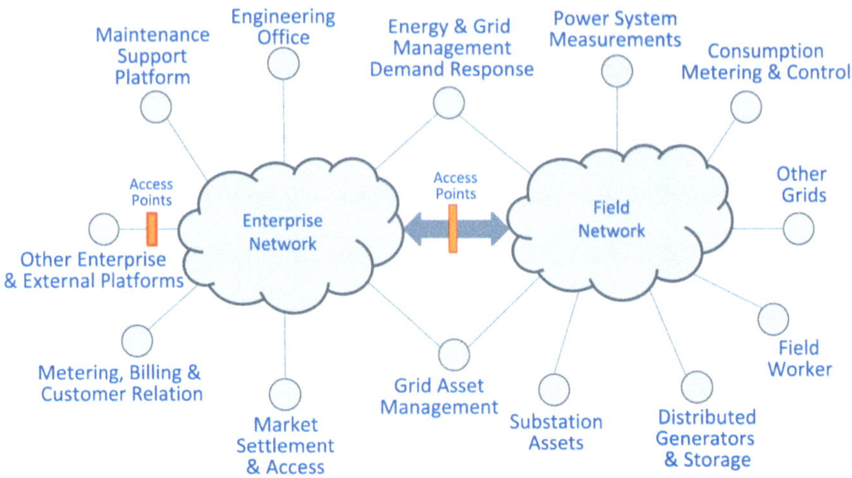

Fig. 6.1 IT (enterprise network) and OT (field network) components of the power utility

It should also be noted that some functional entities need to be part of the two networks. For example energy management systems collect substation data through SCADA platform and may be connected to system operators, independent power generators, and other power utilities.

Similarly, field process sites which are by their location part of the field network shall require access to the enterprise applications if some company employees are physically based at these field sites.

Understandably, decision-makers examine the opportunity of unifying the two networks for cost optimization and for information interchange. However, network coverage and structures, performance metrics, security constraints, management processes and operation practices of the IT world do not always match the realities and requirements of the utilities' operational applications with over-cost and/or under-performance being typical consequences. The main issues of focus are fundamentally different in the two networks rendering a converged network a very challenging concept to implement.

Another solution combining network separation and information connectivity can be based upon limited tunneling across both networks and filtering access barriers for information exchange. Engineering office on the enterprise network can have a tunneled access across the field network to connect substation devices (e.g. asset management), while a field worker at an electrical substation can use a tunnel across the field network for access to maintenance support platform on the enterprise network or even outside the company boundaries.

6.1 Remote Access from Office to Grid Device and Information

This class of applications essentially concern non-real-time transfer of power system data captured in the electrical substation to data analysis platforms and engineering staff for the evaluation of events and confirmation of device configurations. They also comprise the transactional interaction of maintenance staff with substation-based intelligent devices for reading status and measurements, and modifying configurations (e.g., remote diagnostics and tele-maintenance).

The data traffic is sporadic: the available throughput determines the volume of exchanged data files and the corresponding time of exchange, constraining in this manner the potential remote interactions. Assuring the cyber security of the remote access is fundamental in these applications. Remote server authentication (e.g., RADIUS), role-based access control (RBAC) at both ends (engineering workstation and substation device) and access logging are some of the security tools to be employed to assure data exchange security. Deploying dedicated data access servers in the substation can in some cases provide part of the secure remote access architecture. Some typical applications of this type are as follows:

- **Substation Asset Management**—collecting asset information into technical office applications for statistics, correlations, life cycle and failure modeling and preventive maintenance
- **Device Parameters and Settings**—data files uploaded on-demand to provide information on the actual configuration of a device and change of settings according to authorization levels
- **Retrieval of Power System Waveform Files**—typically event-triggered fault records generated by a protection device, a fault recorder or a Phasor Measurement Unit (PMU) can be collected in an engineering office for post-incident analysis
- **Event Reports**—typically log files and reports generated by an event recorder or historical system which provides information on the change of state or a sequence of events in devices.

6.2 Field Worker Access to Central Platforms and Applications

Field intervention is a demanding part of the operational activity requiring a relatively important workforce and logistics depending upon the size and geographical dispersion of the grid and upon required intervention times. Apart from remote asset monitoring, diagnostics and maintenance already discussed in the previous section, utilities mitigate this issue by using nonspecialist field intervention workforce in conjunction with specialist support from in-house or external maintenance support platforms. Some communication applications relating to the connection of mobile or fixed staff located in field sites with central platforms, support applications, and centrally located specialists are as follows:

Substation Operational Voice network

Since the early days of power system telecommunications, there has always been a requirement for providing non-dial hotline telephone connections between control rooms of major operational sites and the load dispatch control center. This hotline telephone was used for manually operating the power system through HV substation operators. At present, the network is generally operated automatically using SCADA for receiving measurement values and status and sending commands to substation devices and many substations are no longer permanently manned. However, highly reliable and secure voice communications are still required in emergency situations and for network switching operations at times of disturbance on the system allowing operational staff to communicate quickly and efficiently. A private, highly secure operational telephone system is still needed to provide the required facilities. The voice facilities in the power system are evolving from a conventional voice PBX system into an IP voice network together with substantial opportunities for developing specific features and functions such as operational call archiving, IP mobility, and back up control center transfers.

Mobile Workforce Applications

Electrical power utilities, in particular those in the distribution segment, have been making extensive use of mobile communications in the management and support of their infrastructure over the past thirty years. In addition to traditional voice services for field-based operational workforce, the evolution of working practices is increasingly leading to mobile data networking applications connecting the maintenance staff to their support base, providing on-line documentation, and workshop applications such as spare parts database. In particular, systems can largely benefit from dedicated mobile data applications designed for workforce management, intervention orders, etc. The operational mobile terminal units are evolving toward smartphones and tablet computers for which ruggedized field-proof versions appear on the market. A number of utilities are undertaking the deployment of advanced mobile workforce communications, in particular relative to disaster recovery. Permanent contact systems associated to "wearable terminals" can keep field workers in continuous contact with support personnel and enable the real-time transmission of camera images and speech from the work site to the support base which can then provide precise support through voice interaction using a headset. These dedicated systems can also provide checklists and templates for the elaboration of on-line real-time reporting and hence improve work safety, accident prevention, and work efficiency. The implementation of data-rich mobile communication systems requires indeed the existence of high availability, disaster-resistant, high throughput wireless connectivity.

More than other communication services in the utility, the provision model for mobile workforce communications is often under assessment. There is no doubt that the most economical solution is to use public mobile services which provide a high level of geographical coverage through extensive deployment of base stations and related infrastructure, as well as continuous roll-out of new data services and applications. The power company pays in this case for its communications only, not for the infrastructure providing coverage. However, an essential application of the mobile system is its usage by maintenance teams during power outages and in disaster recovery situations. The usage of the public mobile in this case is severely constrained due to base stations' insufficient power autonomy and severely degraded service accessibility/performance when the network is widely solicited (e.g., during a disaster situation). This aspect is further discussed in Part 2 on communication service provisioning.

On-line Documentation

Documentation is an essential base for efficient management of utility infrastructure. Previously, the site support staff found all information necessary for carrying out their tasks either at the substation (equipment maintenance manuals and schedules, drawings, etc.) or in their own briefcase. Increasingly, an extensive amount of support information for the field intervention staff is available in centralized servers that can be accessed on-line when required. Pictures and video add particularly useful information in the dispersed environment of the power delivery system. These applications require a broadband network in order to meet an acceptable time performance. The introduction of inexpensive GPS equipment and commercial mapping applications makes Geospatial Information Systems (GIS) an important tool for field based maintenance personnel. Connecting to maintenance applications in the substations and downloading accurate maps, pictures and work orders is an enormous pace in terms of work efficiency but necessitates high network throughput due to growing data volumes and appropriately dimensioned ICT infrastructures. On-line documentation is a well identified requirement and an existing facility in many Utilities.

Collaborative Communications

The working process of utility staff based at field sites is closely following that of their colleagues in the enterprise offices and is being rapidly transformed with new IT practices. Collaborative communications covers all applications that use as terminal the enterprise PC of a member of utility staff allowing him to interact with other utility employees and all corporate applications. They are corporate enterprise applications that need to be accessible in a field site and hence tunneled through the field network to access the enterprise network. The following constitute some of the required services:

- Networked office applications (e.g., mail and calendar systems, file transfer, database access, intranet),
- Work-order and ERP solutions (e.g., project control and time registration),
- IP voice service, conferencing facilities, etc.

These applications require the secure extension of the corporate enterprise applications from the branch office to the operational sites, while remaining fully isolated from the operational applications. IT-support may be effectively administered from a corporate central site.

7 Smart Distribution Applications

Communicating applications in the distribution segment of the power system can be considered as being no different from those described in the previous sections: substation-to-substation, substation or field device to a control platform, and field to office. However, a number of specificities need to be taken into account when designing communication solutions.

- Potentially communicating devices and sites are orders of magnitude larger in numbers and generally dispersed over considerably smaller distances.
- Communicating applications are only emerging with little legacy to be taken into account.
- The electrical grid is permanently changing, communication need to be considerably more agile.
- Time and availability constraints, although present, are less severe than in the higher level grid.
- Distribution system end points are partially outside utility boundaries, at power consumer or dispersed generator premises.

The boundaries of asset ownership and responsibility in the distribution segment depend largely upon the power delivery model and the partitioning of roles in the deregulated context. In very simple terms, one can distinguish (Fig. 7.1):

- **Customer Communications**—Smart metering, customer relations, and demand response
- **Distribution Grid Automation**—Supervision and control of grid's capability to deliver power, DMS/SCADA, Volt-VAR Control (capacitor bank and tap changer commands), FDIR (fault detect, isolate and service restore)

In some cases, common communications can be used for services in the two segments (e.g., transformer monitoring and customer metering) but the applications

© Springer International Publishing Switzerland 2017

C. Samitier (ed.), *Utility Communication Networks and Services*,
CIGRE Green Books, DOI 10.1007/978-3-319-40283-3_7

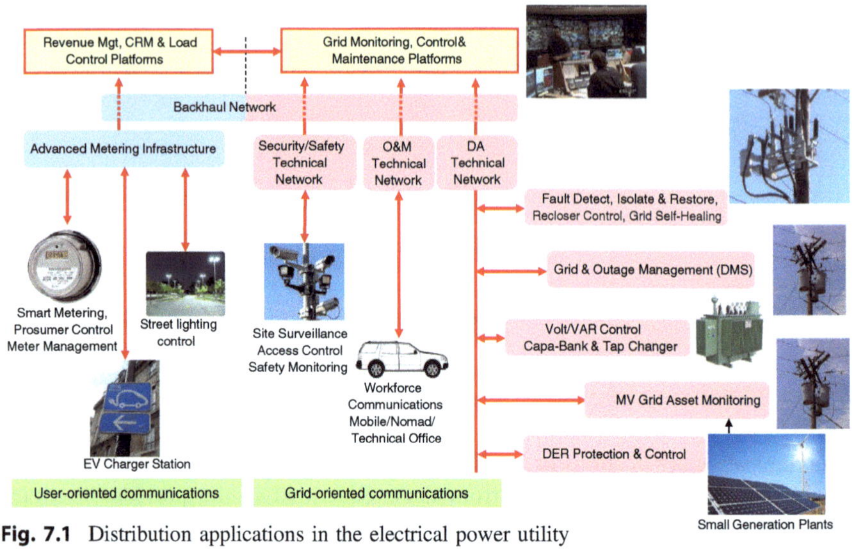

Fig. 7.1 Distribution applications in the electrical power utility

may depend on different companies. The activities may also be separated in future rendering the investments uncertain.

Another distinguishing factor between customer communications and grid automation resides in their tolerance to service outages. Typically metering services can quite easily accept communication service outage consequent to long power outage and can therefore use public communication services. Grid automation, on the other hand, requires extremely high service availability resulting in dedicated communication services.

Common backhauling communications can often be used for both segments using a broadband dedicated telecom infrastructure.

Part II
Provisioning of Utility-Grade Communication Services

8 Service Provisioning, Quality of Service, and SLA

The electrical power utility's decision on the provisioning mode of communication services depends upon several factors the most important of which are listed below:

- Service quality and reliability requirements
- Coverage and service accessibility at sites where communication is required
- Cost of deployment and maintenance of the service during the required operation period
- Commercial availability (or nonavailability) and cost procuring adequate telecom services
- Number and dispersion of sites to cover and their communication traffic
- Company policy and regulatory issues concerning capital and operation expenditure (CAPEX/OPEX)
- Disaster Recovery/Business Continuity and Security constraints
- Company policy and regulatory position on recovering investments through non-operational telecom services (e.g., recovering the cost of optical infrastructure through leasing of dark fibers)
- Organizational issues including in-house availability of skilled staff

Service provisioning can be performed through a public multi-customer operator, in which case the power utility effort is mainly focused on adequately defining the service, contracting the provider's agreement on the quality of the delivered service (Service Level Agreement, SLA) and assuring that the provider delivers the contracted service level.

The power utility can also decide that the most adequate manner to provision the service is to use a dedicated telecom infrastructure through its own service delivery process and organization, in which case the network is generally shared with the utility's other communication services in order to mutualize the cost and effort for service delivery.

© Springer International Publishing Switzerland 2017

C. Samitier (ed.), *Utility Communication Networks and Services*,
CIGRE Green Books, DOI 10.1007/978-3-319-40283-3_8

Different intermediate solutions may exist between the purely procured and fully in-house service provision using different extents of outsourced operation, field intervention or maintenance support. Moreover, services of different kinds or in different parts of the network may be provisioned through different modes.

The performance objectives and the quality of service are also different among these different service types. Many operational services, such as protection relay applications, have extremely severe time delay and communication integrity constraints, whereas the other communication service types are mainly transactional with less severe time sensitivity. On the other hand, business and market communication services implicate access beyond the perimeter of the power company and may raise more severe security issues.

Considering the organizational diversity of electrical power utilities and their different sizes, activities, and regulatory constraints, the exact perimeters can vary to some extent and may evolve with organizational changes. Some of the factors that influence service provisioning are as follows:

- **Security policy**—The definition of separate security domains across the company and the consequent allocation of applications to these different security domains can result in changes on communication service category allocation. This means that the applications which are part of a same security domain shall exclusively use a same group of communication services.
- **Organization**—The organizational entity in charge of a group of applications may require exclusive usage of a service or a same group of communication services.
- **Company strategy**—Grouping of communication services may depend upon the company's strategy, for example to merge corporate and operation-related IT and telecoms, or to merge corporate and market related applications' communications provision, etc.
- **Regulatory issues**—Regulation authorities may prevent operational applications to share communication services with non-operational, or may impose full separation of the commercial U-Telco (Utility-based Commercial Telecom Operator) activities.

When the telecom service providing entity is tightly related to the "Service User" entities, there is a one-to-one correspondence between applications and communication services resulting in an "application-oriented" definition of the communication service (e.g., SCADA or protection circuits). The communication service provider is assumed to be sufficiently familiar with the applications to apply the necessary precautions in the delivery of the required service. However, when a new application is introduced or the requirements of an application change in time, then the user and provider must seek a new common understanding of the service requirements.

On the other hand, where communication service is provided by an external or formally separate entity (e.g., public telecom operator), then the service provision contract defines the service attributes according to the provider's "service catalogue" (interface type, throughput, availability, delay, etc.). The Utility user must in this case decide upon the suitable service category for his application "out of the provider's catalog".

The service provision contract is known as a "Service Level Agreement" (SLA) which can be Explicit in the case of externally provisioned services or Implicit (based on common understanding) for high-intimacy internally provisioned services.

Adopting a particular telecom service provisioning model is not an irrevocable decision. It is often re-examined and reviewed in the light of new situations, some of which are as follows:

1. New company policy and orientation
2. New regulatory issues and requirements
3. Mergers and dislocation of activities
4. Emergence or abandon of adequate telecom services to be procured
5. New applications or change of scale incompatible with the present provisioning model
6. Lack of satisfaction from the services obtained through the existing provisioning mode
7. Major capital investments and running costs required for refurbishment and extension of existing facilities
8. Technological changes in telecommunications and in power system technology
9. Lack of qualified staff and the ageing of the concerned technical work-force.

9 Service Specification Attributes

9.1 Operational Coverage and Topology

Telecom coverage and service access to different operational sites are the first condition for the adoption of any particular service provisioning solution.

- HV Grid sites are accessed in a cost-effective manner through a dedicated telecommunication infrastructure using optical fibers over HV lines (and at another scale using power line carriers). Considering the peripheral location of HV substations and power plants, a telecom operator is not always in a position to provide access with the required capacity.
- Utility offices in urban environment with no proximity to the HV grid cannot be accessed directly through the dedicated network, necessitating the installation of new underground cables, urban microwave, or other wireless "last mile" connections. The approach may be unfeasible or costly. These sites are often more economically served by public telecom operators sharing infrastructure with other potential customers nearby.
- Hydroelectric power generation plants and offshore energy farms often have no other service alternative than dedicated communications or VSAT services. On the other hand, dispersed generation facilities on customer premises may be covered by public telecom operators.

The topology of the network infrastructure also has direct influence on the performance and fault tolerance that can be expected from the communication service (Fig. 9.1):

- The number of transit and switching nodes to be crossed determines the dependability of the connection. Direct links through dedicated fiber over the HV line are preferred for critical applications. Telco services are often discarded in this case due to the impossibility to establish a direct physical link (topology

© Springer International Publishing Switzerland 2017

C. Samitier (ed.), *Utility Communication Networks and Services*,
CIGRE Green Books, DOI 10.1007/978-3-319-40283-3_9

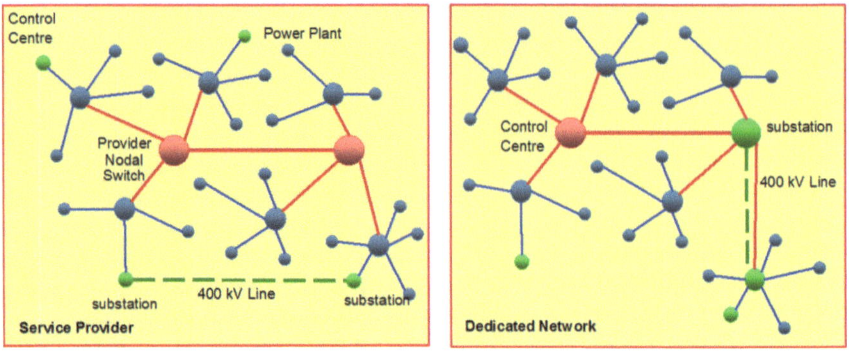

Fig. 9.1 Topology of public telecom provider and dedicated network relative to the power system

is based on criteria such as customer concentration, availability/cost of leased fibers, and site facilities)

- The possibility of establishing two independent routes between two access points of the network, determines the fault tolerance that can be incorporated into the network design. Typically a control center, toward which a great number of communications converge, cannot be located on a secondary spur of the communication network.

9.2 Throughput

Another major criterion for selecting an appropriate provisioning is the potential throughput for different communication solutions. For example, a narrowband HV power line carrier or a UHF wireless system cannot provide the throughput required for a new multi-megabit application.

Throughput is indeed related to an overall communication system capacity which itself is related to the allocated bandwidth across the telecom media. However, throughput being an attribute of a given application, one should examine the sharing scheme of common communication resources among the concurrent applications. A 100 Mbps Ethernet interface does not guarantee a throughput of 100 Mbps to each application using the Ethernet access point. On the other hand, an application generating a sporadic traffic may need high a throughput to evacuate a large data file in a short time and can then liberate the available bandwidth for transporting other traffic, while a small throughput application such as a protection relay may need to transfer measurement samples in a continuous and permanent manner.

Protocol overhead is also a major issue when assessing data throughput. There is an increasing amount of protocol data exchange in many new applications, IEC61850 being no exception. Estimating necessary data throughput for applications must take into account these expanded communication protocols, or to reduce them over bandwidth-critical communication media. One such example is the IEC61850 communication of protection relays across a band-limited transmission medium using Proxy mode stripping the protocol overhead at one end of the link and adding them again by the remote communication terminal (refer to Fig. 3.6).

9.3 Time Constraints

Deterministic and controlled time behavior for communications of time-sensitive applications is one the major reasons leading utilities to deploy and maintain dedicated telecommunication networks. In these networks, the time performance (as well as availability and fault tolerance) can be adapted to the requirements of each application through an appropriate choice and blending of technologies and proper topological structuring. On the other hand, when public telecom services are employed, time control is rarely part of the Service Level Agreement (SLA) of the service provider. Generally, the service provider cannot commit contractually to anything better than an "average communication delay" and is therefore to be excluded when time-sensitive applications are to be carried.

Time behavior of a communication service can be characterized through a number of different parameters as follows:

Time Latency (delay)

Time latency is the absolute delay introduced by the communication network into an application. Time latency is an important constraint wherever a remote command (or remote information needed for elaborating a command) is to be received in constrained time. Time latency also matters where a bidirectional exchange is to be established with limited waiting time at the receiving end.

- Communication channels used for protection relay applications are particularly sensitive to delay. The time allocated to the transfer of protection information across the communication network depends upon the overall fault clearance time determined by practice and policy of the power system (100–120 ms) and its repartition between protection relay operation, protection signaling, circuit breaker operation, and network latency, leaving generally no more than 8–10 ms (half power cycle) for the latter. The transfer time or network latency is a performance constraint which is equally applicable to both analog and state comparison schemes based on system operation times.
- SCADA system overall performance can be degraded by a high time latency or even made completely inoperable through RTU communication time-outs. It should be noted that the "real-time" requirements of SCADA RTU communications are generally in the range of seconds, as compared to order of magnitude smaller transmission times across a thoroughly designed SCADA Ethernet/IP infrastructure. The main issue here is therefore the number of intermediate nodes in the routing of SCADA information as well as the time for any encapsulation and concatenation.

- Voice communication can be seriously degraded by high time latency (more than 100–150 ms).

Absolute time latency problems may be avoided through appropriate design and proper selection of technologies, constrained usage of store-and-forward and traffic queues for critical applications.

With traditional TDM technologies (e.g., SDH) previously employed by public service providers and still widely employed in dedicated utility networks, assuring low network latency is no issue considering the fact that the communication system has no queuing mechanism and therefore no significant buffering at the ingress port of the network or in transit whatever be the speed of the communication system. Basic deterministic design concepts (fiber length and number of intermediate nodes) are sufficient for meeting latency requirements.

In packet communication networks, widely used by service providers and emerging in utility networks, the network delay is composed of

- Packetization Delay (PD)—Time delay introduced by buffering of data before transmitting them as Ethernet packets (e.g., 64 Bytes). The more data we pack into an Ethernet frame, the higher will be the PD, but the higher will be the bandwidth efficiency. On the other hand, we can pack very little protection data into each Ethernet frame so that it can be expedited faster but in this case, the remaining capacity of the Ethernet frame is wasted and therefore a lot of Ethernet bandwidth is required for each communication service.
- Network Delay (ND)—This is the time taken for a packet to cross the network. ND is composed of signal propagation time (5 μs/km) and intermediate node processing (and buffering). It can often be assumed that intermediate nodes operate at much faster rates with reduced buffering and processing times.
- Jitter Buffer Delay (JBD)—JBD is the absolute delay generated by de-jittering, that is to say absorbing delay variations of the network through a buffer. The size of the buffer (in packets) depends upon the amplitude of delay variation that must be absorbed. No need to say that the larger the buffer, the more it introduces absolute delay.

For delay-critical services delivered over IP, preestablished static routes may be employed to ensure guaranteed end-to-end performance.

Delay Variation (Time Predictability)
Time predictability determines the delay variation of a communication channel. It defines the capability to predict the time delay of the communication network, independently from the traffic load from other services being carried across the network, and whatever being the network's state. Time predictability assigns a probability distribution around a nominal time delay value and therefore a maximum acceptable delay.

Transmission delay variation can affect the performance of current differential relays, which require a constant propagation time to start the synchronization of their clocks. Once a delay variation is detected, the synchronization is blocked until the propagation delay is constant again during a period of time. The aim of this check is to use a reliable propagation time, filtering the transients that occur during a switching in the going and return paths. Even though the synchronization of the clocks is blocked, the line differential relays can continue in operation for some period of time (few seconds) as, just before the channel delay variation, they were synchronized. Some line differential relays measure, during the synchronization process, the drift between the clocks. This data can be used to extend the time the protective relays can operate without synchronization through communications. If delay variations occur frequently the synchronization process will be blocked for a long time resulting in blocking the line differential relays.

Delay Variation in TDM systems can only occur after a route change following an automatic restoration. In a packet environment, however, delay variation is mainly due to the existence of service integration and consequent queuing in the network. Time predictability is achieved by avoiding traffic queues which generate variable service times and by imposing constrained routing (e.g., maximum number of hops, predetermined back up route, etc.). It can be absorbed through a jitter buffer at the expense of extra absolute delay.

Delay Asymmetry (Go/Return Path Asymmetry)

Time coherence among the remote points of a distributed application is sometimes achieved by initiating a remote loop-back and measuring the go-and-return transfer time. Considering that the go-and-return times are equal, the transfer time between two sites is in this manner calculated. This type of transfer time estimation is used in older generation differential protection relays (still widely in use) and also in absolute time clock distribution systems. This renders the systems very sensitive to Go/Return path delay variations. In the case of current differential protection systems, a maximal delay asymmetry of less than 400 μs is often necessary.

Delay asymmetry is controlled through avoiding store-and-forward with traffic queuing and variable service time, and bidirectional route switching (i.e., when a fault is detected in one sense of communication, both directions of communication must switch to a same alternate bidirectional route).

Restoration time

Transporting operational traffic requires a time limit on service restoration following a network fault condition. Some operational services require Hitless Switching (no loss of service during the switch over from a normal configuration to a back up state).

Restoration time depends upon the employed communication technologies and the topological complexity of the network:

- End-to-end alternate route switch-over for each critical service is very fast but not scalable.
- Ring protection across an SDH network (e.g., SNCP protection) can restore in less than 50 ms.

- Ethernet Spanning Tree Protocol (STP) has a convergence time which depends upon the complexity of the network (typically seconds to minutes).
- Routing in an IP network is based on different algorithms which reestablish the routing table in each node after any network configuration or status change. Typically RIP-based routing requires less than one minute for restoring communications while an OSPF-based system can restore in 5–10 s.
- Transport Profile Multi-Protocol Label Switching (MPLS-TP) enables the network to restore service using preestablished alternate paths for each data tunnel and can in this way propose restoration times similar to SDH.

Some design aspects in the dedicated network that strongly impact the time behavior of the communication service are given below:

TDM versus Packet network

The migration from conventional TDM (time division multiplexing) networks toward Ethernet and IP increases considerably the network's bandwidth efficiency (avoid idle bandwidth), flexibility (network interfaces and routing), and resilience. However, it can also be a major source of concern for the control of time behavior. Assembling data packets before transmission and store-and-forward of the packet at each intermediate node causes additional buffering delay which increases with the packet size and with the number of queues. Dynamic data routing gives rise to delay variation and lack of time predictability. Static SDH-like packet network solutions exist for dedicated network deployment. However, the public telecom operator packet networks are rather optimized for large multiservice networks where queuing and traffic profiles determine the overall delays.

Multiservice Integration

Bandwidth efficiency in packet networks is achieved through integrating multiple traffic streams into a same packet network. Priority queuing mechanisms are often employed to assure better time performance for more critical services. This implicates that one or multiple traffic queues are established at each routing/switching node. A queue-based store and forward communication system provides essentially a "Best Effort" service with statistical time characteristics. This issue is presently masked through over-dimensioning of the network and "order of magnitude" smaller time requirements of the current applications.

Network Resilience versus Fixed Routing

Network resilience ensures the continuity of service in presence of network faults, but at the same time renders indeterminate the routing of communications. Time predictability is generally sacrificed for improved resilience. Traffic streams which are sensitive to delay variations must generally be treated separately with fixed routing.

L1/L2/L3 Partitioning and Topological Structuring

In order to provide adequate time performance to critical services while maintaining bandwidth efficiency, flexibility, cost, and resilience, it is necessary to design the network with an adequate level of "information forwarding" at physical, data link, and network layers.

- Direct or TDM connections for best time performance but at low bandwidth efficiency (except for services requiring continuous data flows like SV), flexibility, and resilience.
- Ethernet Switching with Virtual Networking (VLAN) and priority assignment for fast transfer of information packets.
- IP Routing for maximum resilience and multiservicing.

Different network topologies can in this way be obtained for different services over a same telecom infrastructure leading to different numbers of intermediate nodes at each layer.

Network Monitoring, SLA Management and Planning

If procured services are to be used for time-sensitive applications, it is important that the Service Provider be contractually committed through SLA to assure the time constraints (absolute time latency delay variations and restoration times) knowing that very often the SLA is not explicit enough to allow many critical applications. Moreover, a contractual commitment on time performance must be continuously (or periodically) monitored and effectively sanctioned: Time performance monitoring functions must be associated to the network's performance management facilities in order to assure that contractual obligations are met.

9.4 Service Integrity and Data Loss

Service integrity is the aptitude of the communication network to deliver the transmitted information without degradation, without loss, and without on-purpose alteration. The present section deals with degradation and loss of information due to channel impairments, on-purpose alteration being covered under communication security further in the section.

In digital communication networks, data channel integrity is characterized by its error performance. This is the capability of the network to deliver error-free data between network's user interfaces. Error performance depends upon the error generating behavior of the network, generally characterized by an average Bit Error Rate (BER) measured over a 24 h time period (excluding 1 s intervals where the BER is worse than 10^{-3}).

Long-term average BER, however, does not contain any information regarding the distribution of errors in time and is hence meaningful only if we assume a fully random distribution of errors (often far from being the case). It is therefore useful to specify different thresholds of error behavior (e.g., BER) measured over shorter time intervals and then, to characterize the communication channel through the statistical distribution of these error behaviors.

Three parameters are defined by ITU-T G.821 to describe the error performance of a 64 kbps connection:

1. **Errored Second (ES)** is a 1-s period that contains one or more errors. For data services the information is commonly transmitted in blocks containing error detection mechanisms. Blocks received with one or more transmission errors are subject to retransmission. In order to have a high throughput, it is necessary to minimize the number of errored blocks. The 1-s period was historically adopted as a compromise value for data block size.
2. **Severely Errored Second (SES)** is a 1-s period where the short-term BER evaluated over one second exceeds 10^{-3}. An SES can lead to a loss of synchronization and it is considered that the connection is unusable during the time interval.
3. **Degraded Minute (DM)** is a one-minute period where the short-term error ratio exceeds 10^{-6}. The degraded minute has been devised principally for digital telephony for which the mentioned error rate is the subjectively perceived boundary of virtually unimpaired transmission.

These parameters are to be derived from averaged data over at least a one-month period. However, in practice, channels are tested for 4 h to assess ITU-T G.821 compliance.

CIGRE used the above-mentioned ITU-T definitions to set a power utility objective on the parameters fixed at 15 % of international connection objectives (Fig. 9.2). It considered a reference connection composed of 5 hops between a remote substation and a control center, and 20 hops in inter-control center connections.

The long term BER value used by ITU-T has been 10^{-6} corresponding to the normal operation of copper and radio communications while the general migration of the system to optical fibers has made 10^{-8} (or even less) a more appropriate long-term BER objective.

ITU-T defined the time limit between unavailability and degraded performance as 10 consecutive seconds. This means that if the bit error ratio exceeds 10^{-3} (SES) for more than ten consecutive seconds, the connection is considered as unavailable for that time. Otherwise, the time interval is considered as available but with degraded performance as shown in Fig. 9.3. Unavailable time begins when ten consecutive SES are observed. It ends when no SES is observed during ten consecutive seconds. These latter seconds count as available time. The counting of degraded minutes is carried out only when the connection is available (excludes unavailable periods).

End-to-end Error Performance Objective for 64kbps Channel		
	ITU-T G821	CIGRE
Errored Second (ES)	8 %	1.2 %
Severely Errored Second (SES)	0.2 %	0.03 %
Degraded Minutes (DM)	10 %	1.5 %

Fig. 9.2 ITU-T and CIGRE 35 error performance objectives for a digital circuit

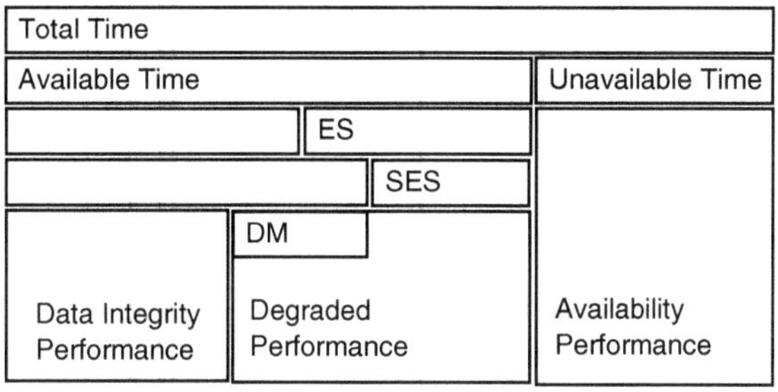

Fig. 9.3 Available time definitions

Despite important technological changes in telecommunications, the error performance objective and its related definitions are still widely used in power utility networks, in particular for planning and testing 2 Mbps connectivity through SDH infrastructure and the primary access multiplexing systems.

However, this can be fully inadequate in some situations. In particular, for critical applications such as Protection Relay communication, "available time with degraded performance" is not a reasonable definition. A system which presents one SES every 10 s (or even 1 SES every 2 s!) cannot be considered as "available with degraded performance" for these critical applications.

Packet Loss and Residual Errors

In packet mode communications, error detection coding and possible retransmission mechanisms prevent the great majority of transmission errors to be seen by the Service User. Impaired information packets are either lost (e.g., in UDP/IP) or corrected through retransmission (e.g., TCP/IP). A marginal amount of transmission errors, called *residual errors*, are undetected by the error detection mechanism and handed over to the user. Operational applications may have a specified objective for the *Residual Error Probability* (e.g., 10^{-12} whatever be the channel BER for control commands) in particular where a residual error cannot be detected at higher layers of the information exchange system or by the application.

Furthermore, an information packet that arrives with a delay outside the application time constraints is considered as lost information. In this way, where a retransmission mechanism exists, the application time tolerance can still lead the valid packet to be considered as lost.

The most common integrity check in a packet data network is therefore Lost Packets statistics through a simple "echo response" (Ping) across the network. Ping commands measure the round-trip time, record any packet loss, and print a statistical summary of the echo response packets received, together with minimum, mean, and maximum round trip times. The command can be of different lengths (number of bytes of accompanied data) to simulate different typical packet lengths corresponding to an application.

9.5 Availability and Dependability

Availability is a service-related statistical parameter which can be defined as the probability of proper operation of the network for a given information exchange. It is normally quoted as a percentage of "up" time of the service, or the percentage of time that the network can effectively forward traffic (e.g., 99.999 %). It can be estimated theoretically and measured practically on a network-wide or per-circuit basis.

Service Availability can be expressed as

$$A_{\mathrm{Service}} = 1 - \left(\frac{\text{Mean Time To Restore Service}}{\text{Mean Time Between Service Outages}}\right)$$

Some currently used values are given below:

Availability objective	Service down-time	Example of service
99.999 %	5.25 min/year (~5 h/57 years)	Protection communication
99.99 %	52.5 min/year (~5 h/5.7 years)	SCADA, operational voice
99.9 %	525 min/year (~5 h/0.57 years)	Data service

However, this statistical parameter, widely used in public telecommunications, in computer systems and networks, and in defining SLAs must be used with precaution when applied to operation-critical services with an extremely low "service-to-idle" time ratio.

As an example, consider a managed service with an apparently high contractual availability of 99.999 %. This gives an unavailability figure of 1E-5, and an unavailable time of 311 s per year.

A Protection Relay system sending 50 trip commands in one year can be "unavailable" during 6 s each time a trip must be transmitted and still respect 99.999 % availability!

For low duty cycle operation-critical services, it is more appropriate to use "**service dependability**" defined as the conditional probability of a service being available when it is solicited. If we consider the unavailability of the service as being independent from (uncorrelated with) the probability of requiring the service, then:

Dependability = Availability × Service-to-Idle Time Ratio
(Considering a service time of 50 times 10 ms per year, the service to idle time ratio: 10^{-8} and the dependability of the service is therefore 10^{-13}).

However, the hypothesis of service unavailability being uncorrelated with service requirement is often far from being evident for operational communications. Typically, a power system fault is a situation that initiates extensive exchange of information, but it also induces impairments in the operation of the communication system.

Unavailability of the protection communication channel can be caused by

A. Excessive communication errors due to channel impairments and synchronization loss (in the network or at the user interface)
B. Network failure, automatic network reconfiguration, and service restoration works
C. Traffic, queuing, and priority impairments leading to unsuitable Quality of Service and late delivery of information
D. Service interruption due to works and maintenance in the network without adequate measures to assure service continuity.

Error-Related Unavailability

As already mentioned, the boundary between unavailable time and degraded performance (available) Time must be set according to service requirements (e.g., through specific monitoring).

ITU-T G821 declares the channel unavailable due to excessive error, if 10 consecutive SES (severely errored seconds) are detected, and available again when no SES is detected for 10 consecutive seconds. This means that the system may remain "available" during 10 s with an error rate that exceeds 10^{-3}, and once declared unavailable, may remain blocked for 10 s even if the channel is error-free. The consequences on security (spurious operation due to non-detected errors) and dependability (channel being non-operating for 10 s intervals) in a protection scheme can be enormous. Availability for protection should be defined in the millisecond range!

In order to avoid undetected invalid frames which may generate spurious (unwanted) operation, the communication channel must be blocked after N consecutive invalid frames. Frames shall then be examined for validity but not used for system operation, until M consecutive valid frames are detected. The communication channel shall be considered as unavailable during this time interval.

Fault-Related Unavailability

Fault-related unavailability is minimized by:

- using more robust, reliable, and fault-tolerant network devices and resources to reduce the occurrence of faults

- associating multiple end-to-end communication routes and media for a more resilient service
- suitable link design, synchronization planning, performance planning, and installation practice
- avoiding too much complexity for faster detection, diagnostics, and repair of network faults impacting communication services and hence for reducing service down-time
- providing prompt detection and correction of faults to reduce the Mean Time To Restore (MTTR). This essential subject is covered in Part 5 on network fault management and maintenance.

QoS-Related Unavailability

QoS-related unavailability is to be avoided through proper traffic engineering and deterministic behavior of the communication network.

Works-Related Unavailability

A major cause of service unavailability in a network can be the inadequacy of service and infrastructure management processes and organization. For example, the atypical availability requirements of the protection communications are not necessarily compatible with a management process designed for delivering IT services. This includes maintenance, extension and change planning, software upgrades and patch management, life-cycle management, etc. Well-trained skilled staff and adapted monitoring are required to avoid this type of unavailability.

9.6 Communication Security

Information security is a global issue covering the entire information and communication system. It is treated as a full end-to-end process through an appropriate Security Policy identifying security domains, taking measures to mitigate risks and devising periodic audit schemes to assess the effectiveness. The subject is extensively treated in other CIGRE publications and by NERC CIP standards (North American Electric Reliability Corporation, Critical Infrastructure Protection).

This section only presents the security requirements for the network expected by the Operational Service User so that the telecommunication connectivity shall not compromise the security level that the User's information infrastructure has achieved. It is in particular valid when there is a separation between the telecom Service Provider and the operational application infrastructure.

Security risk mitigation measures for the telecom Service Provider are presented in Fig. 9.4.

a. The physical access point to the communication service must be secured and possibly allow a network access user authentication (e.g., RADIUS server).
b. The connectivity service across the network must be isolated from other services through dedicated physical bandwidth or through separate virtual private

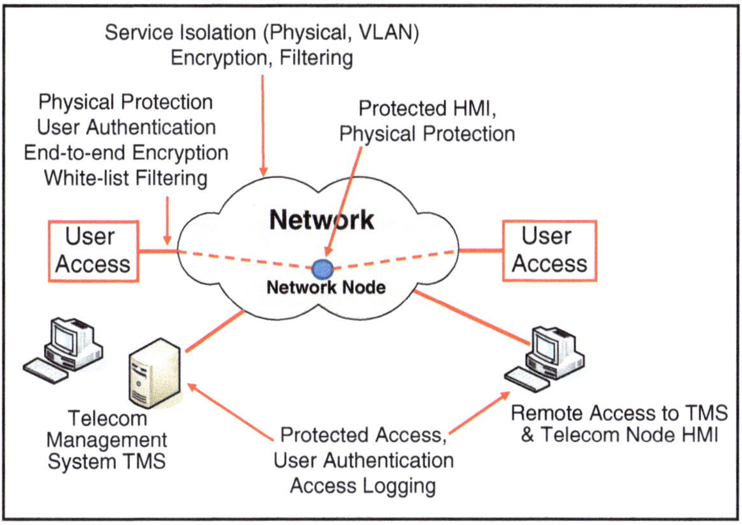

Fig. 9.4 Security risk mitigation measures for the telecom service provider

networks (VLAN/VPN). In case of multiservice IP integration, security barriers and intrusion detection must also be incorporated into the communication network. The connection across the network can further be protected through encryption.

c. Telecom network management platform can constitute a major vulnerability in the provision of communication services. Access to this platform must be secured through physical protection (access to the facilities), authentication and log management.
d. Access to the equipment constituting the nodes of the communication network is also a major source of vulnerability. Restricted physical access to network equipment, secured HMI, and disabling of unused ports and restricting remote configuration and parameter setting are common measures to mitigate risks.

9.7 Future Proofing, Legacy Support, Vendor Independence

Utility operational applications and substation assets have in general a much higher service lifetime than telecommunication services. A newly deployed substation application is expected to operate 10–15 years before being replaced. The adopted communication services for such an application are moreover expected to be "stable and field-proven" at the time of deployment. The communication solution, service, or technology must therefore be sustainable well beyond the expected lifetime of any new generation, mass market, consumer-oriented service or technology.

If a public operator service is employed, the service may disappear before the end-of-life of the power system application. If a dedicated telecom service is used, the interfacing units or equipment may no longer be supported by the manufacturer.

Furthermore, the upgrade of communication system software release or core components, which is a current operation in most communication networks, may be extremely difficult and may require long-term intervention planning if critical applications such as power system protection relaying are to be carried over the network.

Similarly, implementing a new communication infrastructure requires the ability to connect many generations of power system applications rendering the issue of legacy interfacing essential. A new communication solution must provide a way to serve existing applications which may coexist with their replacing systems for a very long time.

In order to assure future upgrade and legacy openness, communication solutions and the corresponding service delivery scheme must not depend upon the use of any proprietary interfaces or technologies and must be as far as possible technology-independent.

9.8 Electromagnetic and Environmental Constraints

Many access points for operational services are at electrical substations, power plants, and other electrical installations. Communication equipment is therefore subject to the same electromagnetic environment as other electronic instrumentation. Service access interface must be adequately protected. Equipment cabinets must be closed, fitted with accessories (earth bar, surge protection, EMC filters, etc.) and wired according to substation installation codes and practices. Cables and wires running within the site perimeter can be a source of conducted disturbances to the communication system. A detailed specification for these EMC immunity aspects is given in IEC61000-6-5 (Electromagnetic compatibility—Immunity for power station and substation environments).

The respect of these precautions and practices is costly and their necessity cannot always be perceived immediately. The consequence of their non-respect can only be checked the painful way during a power system anomaly generating large transient currents and voltages.

The climatic control inside an electrical power site is often minimal and any installed electronic equipment or its cabinet must resist temperature, humidity, dust, etc., at levels which in general do not correspond to telecom and IT environments. Closed equipment cabinets with an adequate degree of protection are generally required. Climatic aspects are also specified in relevant IEC standards.

It should be noted that adopting managed services through a public provider does not remove the expenditure and effort associated to these aspects because access equipment must be installed at utility sites.

Finally, the impact of Earth Potential Rise (EPR) during a station earth fault needs to be taken into account in connecting a new telecommunications service to a HV substation or power station. As an example, an insulation breakdown of a 330 kV asset to earth with a typical fault current of 20 kA may cause the station earth mat to rise up to 8 kV or more above remote earth potential. The exact figure depends on many factors including actual fault current, soil resistivity, earth grid impedance, etc. The way the earth mat, fences, and external connections were initially designed and interconnected would have ensured safety of people on site and remote to site. However, subsequently adding a new telecommunications physical connection without proper understanding of EPR could cause a very dangerous situation to occur to staff at the station or remote from the station due to the difference in earth potential that exists during the earth fault. This problem is solved by avoiding connection of copper communications cables to HV stations preferably replaced by optical fiber or radio solutions. If there is no other economical solution other than connection of a HV zone to a telecommunications Service Provider by a copper cable, then it is essential for safety reasons that appropriate isolation devices are used.

9.9 Service Survivability, Resilience and Disaster Readiness

Survivability is defined as the ability of the communication service to continue during and after a disturbance (network infrastructure fault). Many operation-critical communication services do not tolerate service interruption and down-time. In addition to the statistical concept of availability, it is essential to assure that, as a minimum, no single fault in the reliability chain shall jeopardize the application.

The concept of service survivability perceived at the user end, translates into **Resilience** in the telecom Service Provider's infrastructure. Resilience is the ability to provide and maintain an acceptable level of service in presence of faults.

It is achieved through a multilevel resilience model:

- Fault-tolerant network elements with duplicated core components and duplication of access equipment at critical sites
- Survivable topology (e.g., Ring and Mesh structures allowing alternate routing)
- Disruption-tolerant end-to-end transport through protection switching and service restoration mechanisms (e.g., SDH MSP and Ring protection, Ethernet Spanning Tree Protocol (RSTP), MPLS alternate paths, IP routing mechanisms (e.g., OSPF)
- Fault-Tolerant and adaptive applications and overlays
- Adapted Management strategy and system supervision through appropriate fault management tools dedicated to the operational services.

Some points need to be kept in mind concerning service survivability and the design of resilience:

- **Service Restoration Time**—Different levels of network resilience operate at different time scales which may be compatible or not with different applications' maximum acceptable outage duration.
- **Routing Control**—Designing resilience into the network generally signifies injecting a degree of uncertainty into the routing of information and the resources which are used for delivering the service. This in turn, impacts the absolute time latency of the network and generates delay variation. In the most time-sensitive applications, resilience is restricted to the application itself (main/back up end-to-end routes). The two communication channels for these applications must employ no common path, no common node, and no common equipment, so that no single fault disrupts the application. This requirement requires the controlled routing.
- **Underlying Infrastructure**—Where the telecom service provider network employs contracted lower layer infrastructure, it becomes difficult or impossible to guarantee the independence of main and back up routes. Tracking the full routing of connections can become extremely difficult.
- **Coordination of Resilience**—Applying different layers of resilience in the network to the same service without adequate coordination may cause unnecessary cost and complexity. As an example, main/back up end-to-end application-level fault tolerance, spanning tree protection on the Ethernet connectivity and ring protection in the underlying SDH network may all operate on the same network fault with different time scales to switch the traffic from the same topological path to another unique topological path.
- **Dormant Faults**—An important issue in assuring service survivability is the capability to detect latent faults in the back up routes. A classical strategy adopted in power system telecom networks, is the cross-over of main and back up routes for different applications. This approach uses each of the two paths, and its associated resources, for one main and one back up route, and therefore checks continuously each path to detect any anomalies.

9.10 Cost Considerations

Although the total cost of a service provisioning scheme is not strictly speaking a service specification attribute, it is indeed the major decision criterion once the technical criteria have been used to eliminate "unsuitable" solutions.

Cost assessment for different modes of telecom service provisioning can only be performed in a specific manner for a particular power utility depending upon its existing resources in terms of infrastructure, skills, service procurement opportunities, regulatory issues, and service requirements. Similarly many hybrid provisioning schemes combining service procurement and in-house provisioning may be adopted to fit optimally in any particular case:

- Only particular telecom services may be procured (e.g., mobile services)
- Only particular telecom services may be delivered in-house (e.g., protection relay)
- Access to some particular sites or zones may be through procured services (e.g., urban sites)
- Only some service or layers of infrastructure may be externally provided (e.g., maintenance).

In the present section, we only enumerate some important cost parameters that need to be taken into account when performing such an assessment considering three principal modes of service provisioning which are:

- Build and operate dedicated network
- Build a network and contract services
- Procure communication services.

Asset Ownership Investment Issues on the Regulatory Side

The costs associated with the provision of telecom services can widely be classified into Capital Expenditure (CAPEX) and Operation Expenditure (OPEX). The former comprise the costs associated to the implementation of a telecom network infrastructure, and the latter to the running and maintaining of a telecom network or to the procurement of services. Regulatory constraints in the deregulated environment generally render high CAPEX and low OPEX more attractive because the expenditure on infrastructure may be declared to the regulator as an improvement of the reliability and security of the power system and hence incorporated into the pricing of the service. It should, however, be borne in mind that a favorable regulatory environment can end up to be an obstacle to the integration of nonoperational services (and/or commercial revenue generation through external customers) across the same infrastructure.

Similarly, the rehabilitation of the operational telecommunication facilities can be eligible for financing through international funds which indeed may not be used for procuring externally provided services.

Perimeter of Communication Services and Requirements

When comparing costs and solutions, defining the perimeter of services and requirements is often a tricky issue:

- When provisioning through procured services is envisaged, utilities tend to limit their requirements to the minimal operational services currently in use. However, when planning to implement a dedicated infrastructure, utilities dimension the system not only for present services but also for estimated new site extensions, new applications, and estimated rise in application bandwidth usage during the system lifetime.

- Nonoperational services are generally not included in procured service cost estimations although these services result in some cost to the utility, usually for another department such as "Deregulated Business" department, or alternatively these are "absorbed" by a project or maintenance group so that the true costs are not known. These are most often planned into the dimensioning of dedicated networks.
- There are often auxiliary applications and services that would be implemented if communications were available but rarely deployed when supplementary channels need to be procured.
- When assessing procured service costs, it is usually the "commercially available service" nearest to the required quality which is taken into consideration. There is often an important gap with respect to the actual requirement meaning that the utility accepts to lower its operational expectations. The high cost of actually meeting the operational constraints for atypical services (if possible at all) is rarely taken into account. In the "build" scenarios, however, it is considered that the operational expectation cannot be lowered as technical solutions for meeting the constraints generally exist.
- It should also be noted that quality constraints may be related to the proper operation of a system (e.g., time latency in protection relays), related to the respect of certain national regulations or international recommendations (e.g., fault tolerance), or purely cultural, related to the behavior of the previously employed systems or "how we thought they were behaving" (e.g., invulnerable)!

Capital Expenditure

The cost assessment for the telecom infrastructure needs to be complemented with information concerning the expected useful lifetime of the system components. Optical cable can be expected to have a lifetime of 15–25 years, an RF tower even longer, but modern communication equipment may need to be replaced due to unavailability of spare parts or incompatibility with (or performance limitation of) more recent system components. The addition of varying life factors complicates the cost assessment even more when we consider a procured service counterpart: how to estimate the cost of a currently available commercial service in 15–25 years' time or the cost of adapting the operational applications to whatever telecom service is available in future?

Moreover, it should be noted that the cost of optical cable, RF towers, and electronic communication equipment does not cover all the CAPEX of a "Build" project. Some other important expenditure items often under-estimated are as follows:

- Installation wiring, cable trays, distribution panels,
- Licenses and right of ways
- Power supply, dc distribution, batteries, back up generators, fuel tanks

- Cubicles and environment conditioning, EMC protection,
- Management Facilities
- Application interfaces and converters
- Cost of outage times where critical assets such as transmission lines need to be taken out of service for the installation of OPGW for example
- Spares.

Many of these items still need to be provided in a procured service provisioning scheme: Customer premise equipment in many public telecom operators do not fulfill power system interface requirements and is not conditioned for installation in the substation environment, necessitating cost provision for interface converters, multiplexers, routers, and switches as well as cubicles, DC distribution (or AC UPS for customer premises equipment) and installation items in an assessment. Moreover, if the provider's scope ends outside the perimeter of the electrical substation (e.g., for safety and security reasons), one may also need to add shelters and links to the substation interfacing points. Alternatively the service provider will need to include the cost of gaining and maintaining site access accreditation where he does need to access electrical substations.

Operation Expenditure, Management, and Operation Support

Running a dedicated telecommunication network is costly and requires different categories of skilled workforce, organization, processes, and tools.

Network Management information systems and tools, as other IT systems in the Utility, can be particularly costly to acquire and to maintain. For some more complex management tools, their acquisition may require a certain critical size of the telecom facilities to be managed.

Similarly, field maintenance of a network dispersed across a wide geographical area, while respecting service restoration time constraints can lead to a great number of local maintenance centers with skilled staff and spare parts, but under-utilized most of the time. The operational cost can rapidly become too high to sustain. Sharing field maintenance through multi-skill utility staff (SCADA, protection, substation, etc.) or through external service contractors can considerably reduce this cost (i.e., build and contract services).

However, it should be noted that even a full procurement of "managed telecom services" does not eliminate the cost of management and operation: the external provider relieves the utility from the process of Telecom Resource Management (i.e., the telecom infrastructure) but still the following costs must be taken into account when provisioning solutions are compared:

- Cost of Service Provider Relationship Management—assuring that the provider's contractual obligations in terms of quality of service are met (service level management)
- Cost of Service Management—adapting provider's services to utility user's service requirements, e.g., through coupling of services from different providers

to meet fault tolerance and service restoration times that a single provider can assure
- Cost of Utility Customer/User Relationship Management—assuring that the user is receiving a satisfactory telecom service.

It is a current mistake to assume that the external telecom Service Provider shall replace the utility in the fulfillment of the above-mentioned tasks. This often leaves an important gap in the service provision chain leading to an unsatisfactory service at the user level and conflict with the external Service Provider.

A certain degree of cost sharing on operation support and management can be devised in multinational utilities, covering utility activities in several countries. This can typically cover a centralized support scheme, common procurement, and other administrative activities, however, often constrained by different national regulations and legislative issues, as well as different optimal technology and provisioning schemes.

Skill-Related Costs

A cost item which is so often neglected when performing cost assessments is the cost of maintaining in-house skills and expertise. A first-level superficial analysis may consider that contracting of services (management, maintenance, full procurement of bulk capacity, or process connectivity) leads to important cost saving in that the utility no longer needs to maintain in-house skills and skilled workforce. However, experience shows that losing skills in the utility generally leads to the build-up of a "provision monopoly" for one contractor (e.g., Telecom Service Provider), rise of service cost and degradation of service, due to the inability of the Utility to change its service provider (all the knowledge of the service residing with the external contractor to be replaced).

In some other circumstances, outsourcing of services is not for saving on skilled workforce, but a consequence of lack of skilled workforce or the inability to capture or to maintain skilled workforce (e.g., due to non-attractive salaries).

Finally, it should also be noted that maintaining in-house telecom skills is costly and may even be more costly if the maintained skills are only to be used for specifying and supervising purposes. An adequate provision of operation cost is to be allocated for training, transfer of knowledge between generations and acquisition of new skills facing technological migration.

Risk-Related Costs

Cost assessment of telecom services often lacks consideration of the risk parameter even if this is precisely the reason to envisage anything else than the normal operator service as described throughout this document.

A "Cost versus Risk" assessment must examine the "Liability chain" from the end-user to the end supplier and contractor. The level of liability that any external telecom supplier is willing to accept is completely out-of-proportion with the risk that the utility may assume if the telecom service were not to be operational at critical times. It is often recommended to set the liability of the telecom operator as high as possible (despite the impact on the service price), not to cover the consequences, but to assure responsiveness.

The cost of non-performance of the telecom service comprises the cost of possible damage to power utility assets, cost related to sanctions and penalties that the utility may have to face, as well as the loss of image for lack of performance and its associated costs in the competitive market environment.

Cost Comparison and Selection of the Solution

In order to properly analyze different solutions which have different CAPEX and OPEX cost considerations, it is useful to use appropriate economic models that enable valid cost comparisons to be made. One such standard economic technique is known as "Net Present Value Analysis". In principle this technique converts a stream of future payments (OPEX costs such as continuing telecom service fees, maintenance expenses, ongoing management expenses, etc.) into the present (taking into account the time value of money caused by inflation and opportunity costs) so that they can be added to CAPEX costs. This results in the ability to directly compare different solutions with different mixes of CAPEX and OPEX costs. It is also possible to do a sensitivity analysis for different scenarios, such as changes to the inflation rate during the lifetime of the telecom service delivery. Ample references exist to this technique which can be easily implemented using a spreadsheet approach.

10 Building and Adjusting Service Level Agreements

Whichever the mode of provisioning of telecom services in the Electrical Power Utility (EPU), and the relationship between the service user and provider (formal, semi-formal, or implicit) it is essential to assure a common understanding of the qualities and attributes of the delivered service. The contractual document that reflects these attributes as well as the obligations and liabilities of the service provider toward the service user is the Service Level Agreement (SLA).

An SLA allows the service user to express the operational constraints of its application as defined in the previous sections to the telecom service provider and to obtain the provider's assurance that the delivered service shall meet these requirements.

An SLA allows also the service provider to define the network resources and management processes that he must use in order to meet his contractual obligations towards the service user. Furthermore, the service provider may use the SLA toward his service customers in order to specify the level of service that he expects from his contractors and providers (e.g., underlying infrastructure or support services).

Finally the SLA allows the service provider to know what obligations the service user must meet so that the service can be delivered and maintained by the service provider. Examples may include the provision of racks or floor space for the service provider's equipment, the provision of AC or DC power, access during and out of office hours, third party insurance coverage, etc.

The precision and the exhaustiveness of the SLA become particularly important when the provider is multi-customer and multi-service and the more we move toward a fully procured telecom service.

© Springer International Publishing Switzerland 2017

C. Samitier (ed.), *Utility Communication Networks and Services*,
CIGRE Green Books, DOI 10.1007/978-3-319-40283-3_10

SLA Parameter	Description / Comments
Interface type	As required by the application, (e.g. Optical Ethernet, G703, RS232). Choosing a physical interface such as Ethernet that can be scaled remotely results in easier expansion of services as the need grows.
Bandwidth and throughput	Guaranteed minimum and Peak bandwidth available to the service and degree of flexibility
% number of packets allowed per service for each procured Quality of Service (QOS) level.	It is important to set a policy at the edge of the EPU network to avoid exceeding the allowable limits, otherwise the policy on entry to the Service Provider's network with either drop the packets or remark them to the least priority service, leading to poor service performance due to oversubscription of the service by the EPU itself. Conversely you want to see the Service Provider to apply limiting policies on entry to their network in order to protect the EPU service from contention due to oversubscribed services from other Service Provider customers.
Time Latency (end-to-end delay)	For packet based services, these need to be defined for each class of service. Voice services for example will be processed via separate low latency queues.
Delay Variation (Jitter)	For packet based services these need to be defined for each QOS level. As is the case with most data service parameters these are usually expressed by the Service Provider as monthly averages. Consider how to manage the situation of high peaks that don't cause the monthly averages to exceed the Service Provider specifications. (High peak jitters can cause voice degradation or network convergence problems and still not hit the monthly average parameters.)
Go-Return delay difference	For certain protection relay communications. Asymmetrical delay will cause certain protection schemes to fail.
Service Restore Time on network change	The time required for automatic reconfiguration mechanisms to act upon the network and hence to restore service (e.g. Spanning Tree Protocol, SDH Ring Protection restore time, etc.)
Availability	Distribution, frequency, duration, and timing of service failures.
Integrity and Packet Loss	Specified for each procured class of service.
Power Faults Correlation	Critical services not impacted by power system disturbances. Precautions for not losing service during disturbance.
Resilience and Routing Control	Control of the provider on the routes taken by services in normal time and on anomalies (determines the capability of establishing duplicated communications without common point of failure). The Service Provider and Service User need to agree on the routing protocol between their networks, and to set various metrics that impact on the resilience of the interconnected networks.
Power Autonomy	The time duration for which the service can be delivered in case of A.C. power outage
Maximum Time to Restore Service	Service Provider's ability to respond to service failures and carry out the necessary repairs within the maximum specified time. Different times will be defined for urban, regional, rural and remote locations depending on the location of Service Provider maintenance staff.

Fig. 10.1 SLA checklist for EPU procuring telecom connectivity services

SLA Parameter	Description / Comments
Dual Route Independence	Ability to guarantee that specified connections between two points never use a same equipment, cable segment, power supply, or cable conduit.
Physical redundancy check	Specify the level of redundancy required for example at network level, equipment level, or at specific locations.
Service Isolation & Security	Isolation between internal and external traffic, as well as between different internal services. Measures deployed by the provider to protect against the risks of interfering third parties (confidentiality, denial of service, integrity of information). An EPU will usually have to regard a Service Provider as "untrusted" and employ security techniques such as encryption.
Access Arrangements	Most EPUs have special rules for site access for security and safety reasons. These need to be communicated to the Service Provider and factored into his support of the service.
Qualified/ Certified/ Insured Workforce	Ensure that the Service Provider has sufficient depth in its workforce with the right number of personnel in the right locations to ensure that response time guarantees are realistic. Ensure appropriate insurances are in place to cover accidents by the Service Provider workforce when attending an EPU site.
Performance Reports / Fault Notification	Meaningful and comprehensible information to be provided in a timely fashion. An EPU should consider implementing their own monitoring tools to ensure the performance of the Services is appropriate. This is especially important for packet based services using different QOS levels.
Penalties and Liability	While penalties may not compensate for loss of critical services, they do focus a Service Provider's attention on the need to accurately monitor the SLA guarantees. Usually a Service Provider will exclude responsibility for contingent liabilities and cap their overall liability to a percentage rebate of fees paid. It is worth considering inserting a termination clause in the SLA that allows termination of the service for a sustained poor performance. At least this enables an EPU to engage a new Service Provider and potentially fix the problem using a different service if the current Service Provider continues not to remedy the problem.
Other Legal Conditions	Depending on the structure of the contracts (e.g. if there is a separate service provision contract or not) there may be other legal conditions that may need to be covered off in the SLA including details for; confidentiality between the parties, intellectual property, compliance with all applicable laws (and governing law where the service is provided cross jurisdiction), acceptance and payment, Force Majeure and Termination of contract provisions to name the most common ones.

Fig. 10.1 (continued)

However, very often multi-customer telecom service providers such as public telecom operators provide a catalog of standard SLAs, none of which may meet the requirements of the EPU. Standard "Operator SLAs" are usually not sufficiently

	1 Lowest Severity	2 Low Severity	3 High Severity	4 Highest Severity
Operational Coverage	Control Centres & Corporate Sites	Plants and stations & Control Platform	Along the grid (e.g. workforce)	Beyond the grid, (Energy farms, customer sites, etc.)
Time Latency	1 – 5 sec Human operator	0.1 – 1 sec	Few cycles (20 - 100 msec)	Fraction of a cycle (5 – 20msec)
Time Predictability, Delay Variation	Seconds	0.1 – 1 sec	10 – 100 msec	1 – 10 msec
Delay Asymmetry (go-return path)	May be through different telecom media	Uncontrolled over the same telecom system	Controlled Routing	Identical path, 200μs
Restoration Time	Few Hours	Few Minutes	Few Seconds	100 msec or less
Availability	99%	99.9%	99.99%	99.999%
Service Survivability & Resilience	Service may be lost in the event of anomalies	Survives one module or one link failure	Survives loss of one node or few links	Survives major system faults & disasters
Security Domain	Public	Un-trusted	In Confidence	Protected
Service Integrity	Lost data recovered (Acknowledge & Retransmission)	Not so sensitive to recurrent data error & loss	Tolerates some data loss	High data integrity is critical
Sustainability, Life-cycle Mgt.	Continuous upgrade (type IT)	Yearly upgrade	Multi-annual upgrade (Planned migration)	Constant over application asset lifetime
Environmental Class	Customer Premises Admin Building Control Centre	Power plant / Substation (Control & Relay Rooms)	HV Grid corridors Proximity of HV	Switch-yard Hydraulic Structure

Fig. 10.2 Constraint severity notation criteria

precise to guarantee the fulfillment of operational constraints as described previously and the service provider may not be prepared to review his entire network's operation mode and operational process to meet one customer's requirements. In this case, assessing the most appropriate SLA of the provider against the operational constraints of the EPU applications allows the estimation of the gap and the risk analysis associated to the potential impact of this gap. The following checklists given in Figs. 10.1, 10.2 and 10.3 have been prepared to serve utilities for specifying or assessing SLAs in the EPU operational context.

	Applications / Requirements	Coverage	Time Latency	Delay Variation	Delay Asymmetry	Restoration Time	Availability	Survivability	Security Domain	Service Integrity	Life-cycle Mgt.	Environment Class
Operational Applications	Protection Communications Current Differential	2	4	4	3-4	4	4	4	4	4	4	2
	Protection Communications State Comparison (command)	2	3	3	3	4	3	4	4	4	4	2
	System-wide Protection (WAP&C)	2	2-3	3	3	4	4	4	4	4	4	2
	Remote substation control	2	2	2	2-3	3	2	3	4	4	3	2
	Operational Telephony	2	2	2	2	3	2	3	4	2	3	2
	SCADA RTU	2	2	2	2	3	2	2	4	3	3	2
	Generation Control Signaling	2	2	1	1	3	2	2	4	4	3	2
	Inter-control centre communication	1	2	1	2	3	2	2	4	1	1	1
	Remote Operator	1	1	1	2	3	2	2	4	1	1	2
	Synchrophasor visualization & monitoring (WAMS)	2	1	1	1	2	1	2	4	2	3	2
	Settlement and Reconciliation metering	2	1	1	1	2	1	1	3	3	3	2
	Smart Metering	4	1	1	1	1	1	1	3	1	3	1

	Applications / Requirements	Coverage	Time Latency	Delay Variation	Delay Asymmetry	Restoration Time	Availability	Survivability	Security Domain	Service Integrity	Life-cycle Mgt.	Environment class
Operation Support	Mobile Workforce Communications	3	2	2	2	2	2	4	1-2	2	1-2	4
	Collaborative Multimedia Comms.	2	2	2	1	1	1	2	2-3	1-2	1	1-2
	Automation Device Management	2	1	1	2	2	1	2	4	3	3-4	2
	Substation Data Retrieval	2	1	1	1	1	1	1	4	1	1-2	2
	On-line Documentation	2	1	1	1	1	1	1	3-4	1	1	2
	Condition Monitoring	2	1	1	1	1	1	1	3	1-2	3-4	2
Security & Safety	Video-surveillance of sites	2	1	1	1	1	1	3	3-4	2	3	4
	Site Access Control	2	1	1	1	1	1	4	3-4	1	3	2
	Environment Hazard Monitoring	2	1	1	1	1	1	4	3-4	1	3	2
	Intruder Detection	2	1	1	1	1	1	4	3-4	1	2-3	3-4
	Isolated Worker Safety	3	1	1	1	1	1	4	3-4	1	3	3-4
	Public Warning Applications	4	1	1	1	1	1	4	3-4	1	3	4
	Hydraulic Stress O&M	2	1	1	1	1	1	4	3-4	1	3	4
	Cyber-security Applications	2	1	1	2	1	1	4	4	1	1-2	2

Fig. 10.3 Typical communication service requirements for power utility applications

Figure 10.3 titled "Typical Communication Service Requirements for Power Utility Applications" provides a cross reference of typical service requirements for utilities' applications. The reader should use Fig. 10.2 for characterizing the severity levels 1, 2, 3, and 4 in Fig. 10.3.

Service Provisioning Models—Impact on the Delivery Process

11

The relationship between the Electrical Power Utility (EPU) operation-related telecom service user and the corresponding telecom service provider can take multiple forms and can also change over time. Figure 11.1 presents schematically the main patterns encountered in the power utilities. It should be noted that in a same EPU we can find different schemes for different groups of services, different layers of telecom service, or different geographical areas. The pattern may change due to EPU change of policy, regulatory changes, or the evolution of technologies.

This section provides some in-sight into the reasons for adopting each and the corresponding issues that may arise.

Integrated to the Operational User (Type A)

This scheme is the most basic and historically the most employed form of telecom service provision in the EPU. It relies upon the total ownership of all telecom assets and in-house provision of skills for running the network which can be designed, deployed and periodically refurbished through turn-key contracts, or gradually created through substation, transmission line and SCADA procurements.

Providing telecom services as an integrated activity of the EPU operations has major advantages which are particularly important where "market-atypical" operations-critical requirements such as those of Protection communications are concerned:

- **Full commitment**—The network specifications in terms of performance, topology, and capability perfectly reflect the user requirements. The telecom staff's priority of the day is the operation staff's current problems.
- **Informal relationship**—Telecom staff are direct colleagues of protection, substation automation and SCADA engineers. Performance issues and interface requirements, intervention scheduling and problem solving do not risk to be compromised due to misunderstanding. Interaction with telecom network management is through internal meetings without immediate need for SLA and contract management.

© Springer International Publishing Switzerland 2017

C. Samitier (ed.), *Utility Communication Networks and Services*,
CIGRE Green Books, DOI 10.1007/978-3-319-40283-3_11

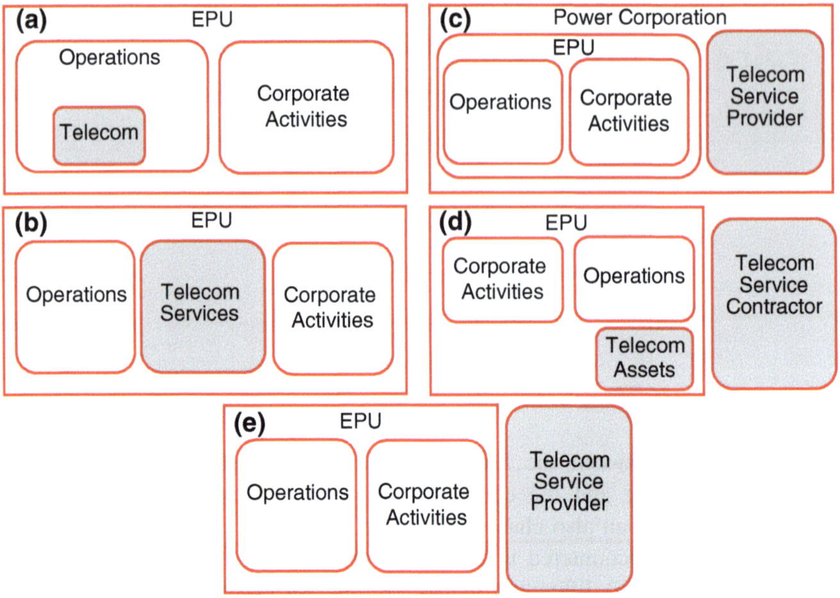

(a) Telecom is part of the operational activity. Corporate entity provisions telecom services separately.
(b) Common Telecom (& IT) Services for both Corporate and Operational Applications.
(c) TSP is a sister company to the EPU, providing services exclusively (or in priority) for the Power System
(d) EPU procures its telecom assets but operates them using an external Service Contractor
(e) Telecom services are procured under SLA by a TSP providing services to many customers.

Fig. 11.1 Telecom service provision models in the EPU

- **Maximal responsiveness**—The intervention time of maintenance staff in case of service interruption is not prolonged due to site access issues and when multiple interventions at application system and telecom level are required, this can be coordinated in minimal time with only internal field staff likely to be based at the same field maintenance center.
- **Synchronized deployment**—Addition or upgrade of telecom services when a new application is deployed or when the power system is extended need not be anticipated long time in advance for provisioning of necessary telecom assets and scheduling of works. Application and communication service can be provisioned together or at least in a synchronized manner.
- **Information Security**—The telecom system and the corresponding organization and processes being an integral part of the EPU operations, they are covered by the same security policy. No coordination action or additional auditing is required to assure that the security policy of the service provider shall not compromise that of the EPU.
- **Disaster Recovery/Business Continuity Planning (DR/BCP)**—As for information security, the telecom organization and processes are an integral part of

the EPU operations. No coordination and additional auditing is necessary to assure that DR/BCP of the provider is not compromising that of the EPU.

The main drawback from this service delivery scheme is indeed the limited possibilities of a constrained telecom team operating inside the EPU operational entity. The team shall be dealing only with the operation-related telecom service requirements of the EPU and shall therefore be unable to implement more complex, more costly, and more demanding technologies, management tools, or at a very high cost due to the small scale of the requirements.

Another particular concern for this model is its lack of performance and efficiency measurement through SLA and cost prospective. The quality and cost of the delivered service is not truly assessed against any particular reference.

An integrated telecom service provision scheme can scale up to cover corporate or other communications inside and outside the EPU, but in this case, the evolution to a type B situation is almost automatic in order to cover assets and running costs for the corporate communications.

Sister Entity to the Operational User (Type B)

The normal position for an "internal" telecom service provider who delivers services to both operation-related and corporate enterprise applications is an entity independent from both. This position allows the delivery of services in a "semi-formal" relationship with a larger traffic volume and Service User base.

The provisioning scheme allows to deploy a core network common to operation-related and corporate services, and to employ data networking and IT specialist skills (necessary for the corporate communications), in order to implement new generation operation-support services.

This scheme is often the "minimum critical mass" necessary for the implementation of "enhanced" network and service management tools.

The internal nature of the telecom service provider still allows a fair level of commitment although not as informal as the type A scheme.

Affiliated Service Company (Type C)

Provision of external services (U-Telco) or simply the intention of creating a separate profit making company can lead to the extraction of the telecom service provider from the utility organization.

A type C scheme is different from type B mainly in its degree of freedom in investment and its consequent overall accountability.

The company can in particular:

- Procure its own new assets or extend their capacity,
- Design new services,
- Extend its customer base to competitive telecom market,
- Employ its needed skills and pay competitive salaries to maintain its staff,
- Subcontract tasks and services to specialized contractors.

The relationship with operation-related organization is more commercial and based on annual negotiations based on SLA or service contracts.

Service management is formal but in most cases, the history of the telecom Service provider (converged in the recent past with that of the operations entity) often allows informal relations and knowledge of the operational applications masking any shortcomings in the formal process. In time, more formal specifications and information exchange processes must replace the "ex-colleague corrective patches".

Service commitment for operation-related services (whether based on SLA or not) remains the high priority and fundamentally different from SLA commitments toward U-Telco customers. In the former case, failing to deliver service may lead to enormous damage at the mother company EPU and in the latter case, only to limited financial sanctions for not meeting an SLA.

The liberty of the company in terms of development strategy, assets, and human resources and extra income from sharing the infrastructure with other users (or providing services to external customers) normally results in a more cost-effective telecom service provision and should lead to lower service costs for the EPU. On the other hand, the telecom service provider must assume the responsibility for network planning, development, and refurbishment of communication network and service platforms in order to maintain the quality of the delivered service (e.g., mitigate asset aging) and to ensure that the infrastructure is capable of responding to new requirements (new services, increased bandwidth requirement, and service migration) provided that the EPU ensure the financing. This requires periodic assessment of EPU migration plans at the time of revision of the service catalog and pricing.

However, delivering U-Telco services can also lead to telecom regulatory issues and in particular fair trade regulations loosening the preferential links with the EPU. Depending on the proportions that external service provision may take in comparison to the EPU service, the danger is that in time, the affiliated telecom company may become simply a normal commercial service supplier resisting the specificities of the EPU's operational services.

Independent Service Contractor (Type D)

An EPU requiring specific telecom services but not intending to maintain the necessary skills and organization, may deploy a dedicated telecom infrastructure and maintain the network by an external contractor.

The perimeter of the service contract may vary according to EPU in-house capabilities:

- Service Management
- Telecom Infrastructure Management
- Field Maintenance

The contractor provides organization, process and skills, even the absorption of EPU's telecom staff and can often better maintain the skilled workforce through more competitive salary policy than the EPU itself. On the other hand, the EPU shall lose technical knowhow in medium/long term and consequently the control of its network and of its contractor.

The contractor is engaged with a Service Level Agreement governing its interventions and services but is not responsible for the failure of aging assets or their lack of performance whose renewal policy remains with the EPU employer even if the contractor conserves an advisory role in this respect. Typically, the service contractor must prepare a yearly development and refurbishment plan of communication network and service platforms based upon the EPU plan for application changes and the contractor's survey of aging assets. The contractor can only assume the responsibility of maintaining the quality of the delivered service if the EPU accepts the refurbishment and new developments ensuring that the infrastructure is capable of delivering the service.

External Telecom Service Provider (Type E)

The least degree of EPU involvement in the delivery of necessary telecom services is to procure it according to an SLA from a multi-customer Telecom Service Provider such as the Public Telecom Operator.

Procuring telecom services liberates the EPU from procuring assets, deploying infrastructures, employing skilled workforce, building processes, and deploying tools for its management and maintenance. However, the EPU shall still need to manage the external service provider with adequate processes (and tools) and adapt the procured communication resources to the requirements of its internal users.

The infrastructure is extended, diversified, upgraded, and renewed without any involvement from the EPU. However, extensions, new services, and service migrations need to be planned long in advance to ensure that the provider shall have the capability of delivering the new services (e.g., covering new sites, increasing capacity in remote areas, etc.). This will be included in the yearly renewal or revision of service contracts.

However, this mode of service provisioning presents many drawbacks which are symmetrically opposite to the advantages given for Type A described above. The EPU will have, in particular, to provide considerable effort in the following domains:

1. Formally and precisely specify service requirements and constraints. It should be noted that the terms and vocabulary do not have the same significance in public telecom and in the operational EPU context (e.g., availability) and may lead to misunderstandings with great consequences. Time behavior and predictability of the connections may be an important point to consider.
2. Establish Service Level Agreements (SLA) and Sanctions for not respecting them—It should be noted that non-respect of SLA in the world of telecom is sanctioned by financial compensation with no proportionality to the EPU risks due to lack of service.
3. Carry out Performance Measurement and SLA Monitoring with appropriate tools.

4. Provide considerable effort in contract and conflict management.
5. Implement application interfacing and service multiplexing in operational sites where the service operator cannot access.
6. Coordinate Security Policy and Disaster Recovery/Business Continuity Plan of the Service Provider with those of the EPU. Perform audits to assure that they are not compromised. In particular, power autonomy, or the capability of the telecom service to be delivered in the event of a power outage through adequately dimensioned batteries is of great importance for disaster recovery.
7. Schedule long in advance any extensions, changes, and upgrades and negotiate in good time with the provider.
8. Avoid monopolies and dominant positions for any single telecom provider which may increase its prices and decrease the quality of service.
9. Service life expectancy has to be carefully analyzed before using extensively a standard service delivered by a provider. Many cases can be enumerated where a standard telecom service used by an EPU is abandoned or replaced by another service not equivalent for EPU usage (e.g., leased digital circuits used for protection relay communications).
10. "Safety certified" field maintenance workforce or "safe location" for provider's assets.

To sum up, no single service provisioning scheme can be considered as optimal in all situations and for all power utilities. As it was stated previously, different telecom service provisioning modes often coexist in the same EPU depending on the nature of services.

- When operation-related telecom services are provisioned through an integrated entity (type A), then corporate communications are generally through procured service (type E).
- When operational and corporate services are integrated into the same provisioning model and organization (type B, C, or D), then protection communications are often separated from this integrated approach and performed directly through separate fibers or wavelengths (type A).

Part III
Delivery of Communication Services in the Utility Environment

12 Introduction on Service Delivery

Service delivery is the direct counterpart of service provisioning: they represent together a customer to provider relationship. How operational applications can be specified in terms of requirements and consequently provisioned as communication services was discussed in the previous part. The present part describes how these specified services can be delivered with the required qualities by a telecom service provider, and in particular, by the utility's in-house telecom infrastructure, staff, and processes.

Delivering communication services at a point of access of the utility applications implies

- appropriate interfaces for adapting to different types of service,
- an adequate way of sharing communication resources between different applications,
- an appropriate network architecture capable of providing different levels of service quality.

However, service delivery is not just a question of having necessary network resources, but also a process describing how services are provisioned, maintained, modified, and monitored in order to satisfy each utility user. The interactions between the telecom service providing organization and the service users are covered in the present part, while the interactions inside the telecom providing organization for operating and maintaining the network resources are covered in Part 5 further in the book.

The cost and quality of services or of the delivery scheme comprising network infrastructures, tools, and staff is continually assessed against other internal or external provisioning options. The scheme must therefore be examined, adapted, and modified in accordance with quantitative and qualitative changes occurring in the communication services that are to be delivered.

This part begins with building an architectural model for the delivery of telecom services, introducing the principles of Application Networks with Wide Area

© Springer International Publishing Switzerland 2017

C. Samitier (ed.), *Utility Communication Networks and Services*,
CIGRE Green Books, DOI 10.1007/978-3-319-40283-3_12

Network connectivity through a Transport Network. The overlay model currently employed in utility telecom networks and its suitability for the slow migration encountered in utility networks is described.

The section continues with a discussion of the delivery process model and its constituents as applied in the Utility context. This is part of the more general Utilities' Telecom Operations Map as defined by CIGRE D2.26 working group which will be further revisited in the corresponding parts on deploying the network and maintaining the network operation.

Then the concept of applications interfaces adaptation and their aggregation at the transport network's edge is discussed. Special focus is given in this section to substation access to broadband communication resources such as optical fibers. Different combinations of service and transport are reviewed associating Ethernet- or legacy-based services with packet- or TDM-based transport.

Service management and service-level monitoring are covered next and finally this part is closed with a review of the issues of aggregating or integrating operational and enterprise services in the power utility context.

Communication Service Delivery Architecture

13

Utility applications requiring communications to be transported across the power system require different network topologies and quality requirements. The ownership and the management of these communicating applications are under the responsibility of different entities in the power utility organization. Moreover, they are not deployed at the same time and cannot be enumerated exhaustively for the life time duration of the communication network. It is therefore common practice to separate **Application Networks** (**AN**) from the **Transport Network** (**TN**) as illustrated in Fig. 13.1.

Each AN (SCADA, voice, etc.) caters one or several applications with similar characteristics (performance, security, topology, ownership, etc.). It can be a point-to-point connection of two substation-based devices, a group of interconnected local area networks (LANs), or a control platform connecting to remote substation devices. The overall mapping between applications and ANs is an essential part of the design and should allow for the future increase and growth of devices on each AN or the addition of new application networks.

The TN is composed of telecom infrastructure managed by an internal or external Telecom Service Provider delivering connection services with appropriate quality to every AN at each site. The TN delivers a transport service to each AN through a provider edge device at a **Service Access Point** (**SAP**). The quality of the delivered services is assured by the technical performances of the network and the way each traffic stream is treated in the aggregated data stream. This consists in allocating communication resources such as bandwidth and also adequate mechanisms for controlling the access to shared resources collectively called quality of service control (e.g., allocating priority in a service queue).

The separation of ANs from the TN allows distinct migration planning in the substation and across the network, as well as simplified network management, quality monitoring, and security management processes. The termination device for the AN/TN and SAP are the contractual points of service delivery, monitoring, and user/provider management.

© Springer International Publishing Switzerland 2017

C. Samitier (ed.), *Utility Communication Networks and Services*,
CIGRE Green Books, DOI 10.1007/978-3-319-40283-3_13

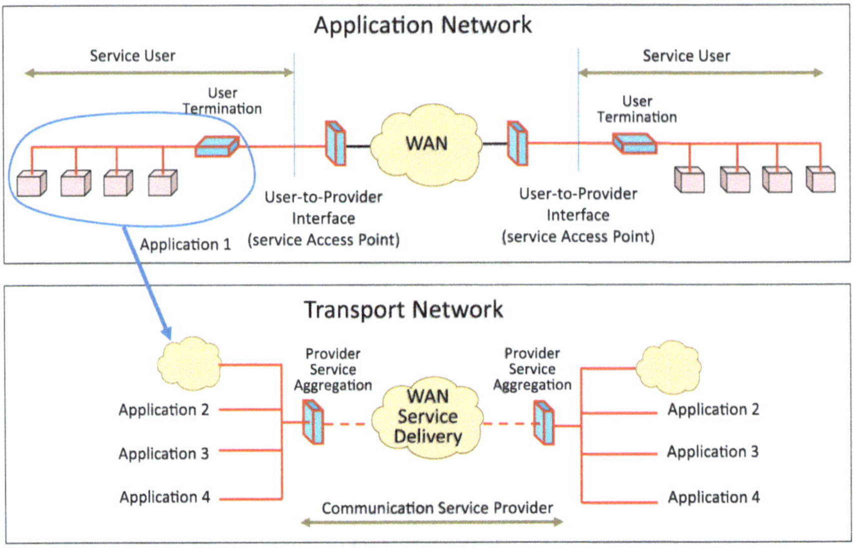

Fig. 13.1 Application networks (AN) and transport network (TN) separation

As shown in Fig. 13.2, converging of data streams from different ANs into a same telecom transport network necessitates service access and aggregation devices at the substation boundary. This device adapts the application interfaces to the transport requirements and manages the repartition of the AN bandwidth between different AN.

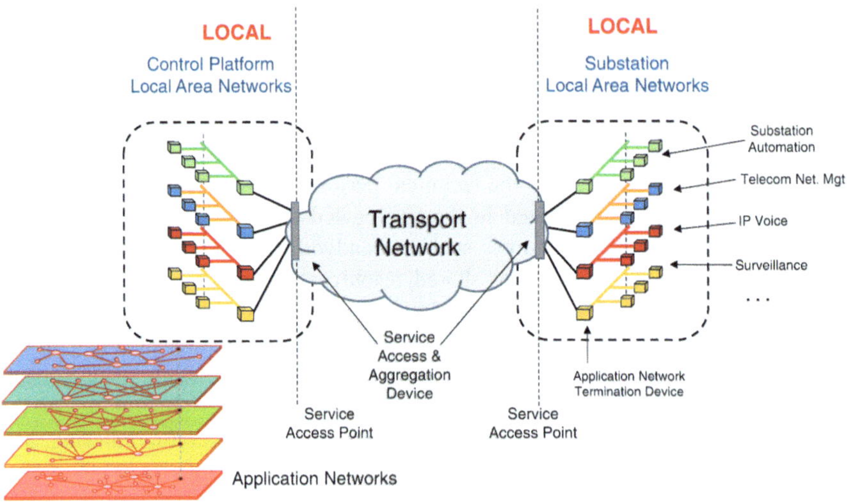

Fig. 13.2 Architectural model for delivering operational communications services

Application networks have different interfacing requirements. Older generation applications in the substation generally provide an analog or digital circuit interface of various types at the SAP. New IP-based applications present natively an Ethernet packet interface to the TN. Interfacing and interface adaptation is discussed further in the Chap. 14 hereafter.

The aggregation of AN data streams can be performed at different levels ranging from distinct fibers or distinct wavelengths to static or dynamic allocation of time for the transport of each data stream. Time allocation can be performed by time division multiplexing (TDM) allocating time slots in a cyclic manner to each data stream or by implementing one or multiple queues at a "store-and-forward" node and hence giving a switched access to network resources for data packets from different application sources. In the former case the device is a TDM access multiplexer and in the latter, a packet switch. It should be noted that all ANs are not necessarily integrated in the same manner and at the same level.

Distinct fibers and static wavelengths are mainly applicable in simple cases and at small scale: their implementation cost and effort as well as their structural rigidity make them appropriate mostly for subtracting one particular data stream to ease the constraints on the network or to respect organizational separation (e.g., separating protection communications or corporate enterprise traffic (IT) from the operational telecommunication (OT) system). Separate fibers or wavelengths can also be employed for hierarchical structuring of the TN as discussed in the following lines. More dynamic usage of wavelength multiplexing is being developed in large TN but still over-dimensioned for operational telecom networks of power utilities.

Full IP integration, at the opposite end, uses a same logical network, routing process, and addressing plan to transport all traffic from different applications. Services are prioritized by their assigned QoS class and packet routing is performed according to the destination address. IP service integration, the general solution in enterprise networks, gives maximum resilience to network change and facility for creation of new network users (i.e., where traffic streams cannot be specified and enumerated in advance). It is to be used where the size and dynamics of the network justify automatic routing change mechanisms at the expense of reduced control over network behavior and performance. This is the case in large distribution grids where thousands of field nodes are to communicate to remote data processing platforms

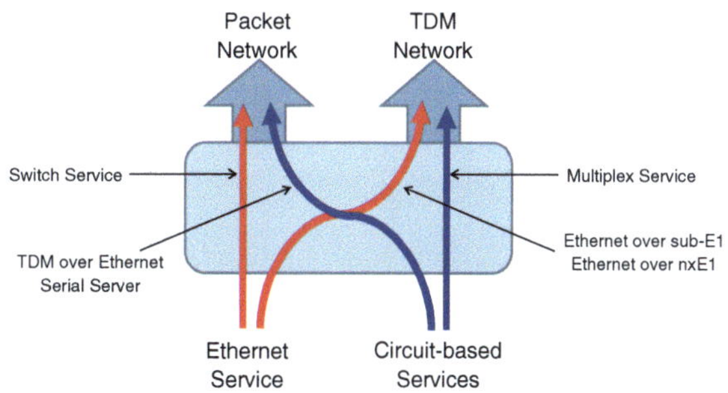

Fig. 13.3 Hybrid packet/circuit service access and aggregation device

often without very tight time imperatives (e.g., smart metering). It is also a valid solution inside some AN, for example for network supervision, remote maintenance, and field worker communications. Cyber security issues are particularly important in this context.

Between the two opposite cases of IP service integration and physically separate networks, application networks can be aggregated in different intermediate manners with advantages and drawbacks which depend largely on the extent of packet-based Ethernet and legacy circuit-based interfaces, access to fixed allocated and shared telecommunication network capacity, and required degree of flexibility. Electronic technology integration increasingly enables the design of hybrid access solutions where packet switching and legacy multiplexing and associated interface adaptations can coexist for gradual service migration as often required in power utility OT networks as shown in Fig. 13.3.

14 Service Interfacing at the Access Point

14.1 Legacy Interfacing

Unlike the office network where device interfaces have been standardized to wired IP/Ethernet since a long time (with wireless increasingly taking over today), the power system communication network interfaces have till recently remained largely diversified. The most common, but far from being exhaustive, are as follows:

- EIA RS232C/ITU-T V.24–V.28—is probably the most widely used serial communication interface in the power system allowing communications of SCADA RTUs, Instrumentation HMI, recorders, data loggers, alarm transmitters, etc. This is a non-balanced interface consisting of 1-wire Tx and Rx signals + Control signals + Ground + clocks for synchronous mode. It can transmit synchronous or asynchronous data at speeds up to 20–30 kbps.
- EIA RS422/RS423 (ITU-T V11/V10) allows higher speeds and longer distances than RS232. Transmit and receive signals are balanced (2-wires) for RS422/V11. It is mainly used for applications requiring higher speed than 20 kbps (e.g., teleprotection signaling).
- ITU-T G.703—is the telecom carrier interface for transporting a full primary multiplex at 2048 kbps, currently referred to as 2 Mbps or E1 (as well as its US equivalent T1). In the power system, it corresponds to an output of the primary interface multiplexer, to an input of the SDH transmission system, or to the communication interface of certain devices requiring high speeds or bypassing of the primary multiplexer. G703 also defines co- and contra-directional 64 kbps interfaces used by some protection signaling and current differential protection relays now being replaced by E1, by IEEE C37.94 or by Ethernet.
- Analog 4-wire Voice Frequency interface(4-wire E&M)—This interface is used not only for connecting legacy analog voice systems now largely replaced, but also as a transparent signal transmission channel for older generation substation devices such as voice frequency modems or protection signaling.

© Springer International Publishing Switzerland 2017

C. Samitier (ed.), *Utility Communication Networks and Services*,
CIGRE Green Books, DOI 10.1007/978-3-319-40283-3_14

- Subscriber-side and Exchange-side analog voice interfaces—these 2-wire interfaces called FXS and FXE provide signaling, loop-disconnect, ringing current, and all other functions necessary for connecting an analog telephone set to a remote telephone exchange.
- IEEE C37.94—is an optical interface allowing the transmission of Nx64 kbps between protection relay and the communication system in the substation local area ($\sim$2 km) over optical fibers and hence avoiding electromagnetic influences and potential rise on copper wire connections.

These legacy interfaces in the electrical power substation are bound to disappear in time but with a lifetime that can be very long. The replacement and re-commissioning of protection relay devices and SCADA RTUs require significant effort and is typically extended in time according to a gradual migration plan. It is therefore important to keep legacy device interface aggregation (i.e., substation access multiplexers) as long as these legacy interfaces persist. Interface conversion can be costly when used in extensive manner, is a source of faults and human errors, and degrades performance.

14.2 Ethernet Access

Replacing legacy data interfaces in the substation by Ethernet ports presents the following benefits:

- Single interface replacing many different functional, electrical, and mechanical interfaces for data and voice circuits are used in the electrical power environment (RS-232, RS-422/V.11, RS-423/V.10, RS-485, X.21, G.703, etc.). This interface standardization results in reduced engineering and coordination effort, reduced documentation requirements, reduced spare parts, and no interface converters.
- Low cost of connection point—RJ45 is the most economical connection point well below any other data interface (connector, patch cord, port hardware, network hardware).
- Bit rate flexibility—All interfaces are at the same bit rate. The throughput at the interface and across the network is soft controlled and can be modified without changing any physical boards or channel reprogramming across the network.
- Fiber isolation—Spans of cable that require galvanic isolation or electromagnetic immunity can be implemented in optical fibers often without any external electro-optical converters.
- No Protocol conversion and transcoding—Knowing that at the central node, generally it is required to connect into a platform on an Ethernet LAN, direct Ethernet interfacing at the substation end avoids protocol conversions.
- Easy implementation of back up control center routing—No junction splitters required for broadcasting data to two different destinations.

- Stable and standard protocol—Ethernet is the most dominant standard in the networking industry. Its widespread use guarantees its availability for a very long time.
- Transmission span—Ethernet allows much longer separation between the application equipment and the network equipment than any of the legacy data interfaces that it replaces.
- Strong industry support—Ethernet technology is available from a large number of suppliers with continuous development (switching, speed, fiber, wireless …) both at the supplier end and at the standardization end.
- Extensive topological flexibility—Through the use of Ethernet switches it is possible to adapt to any topology and environment constraint.

Various standard Ethernet interfaces have been defined and used extensively for different applications. Industrial-grade versions exist abundantly on the market. The most common operate at a line bit rate of 100 Mbps (Fast Ethernet) over copper wire and over optical fibers. 1 Gbps interfaces Gigabit Ethernet (GbE) are mostly used in the network interconnecting communication nodes or inside the substation LAN but not common in the connection of an application network to the transport network (user-to-network interface).

Higher bit rate Ethernet interfaces are continually being developed and introduced in different application domains. 10 G is widely available and occasionally used in utility core networks, 40 and 100 G for the moment only in mainstream telecommunications.

15 Synchronization at User-to-Network Interface

Clock synchronization and time distribution are important issues and potential source of system anomalies at the Service Access Point (user-to-network interface). For analog and asynchronous data interfaces there is no synchronization issue: the user and the network worlds are fully isolated. However, when dealing with synchronous data interfaces, a rigorous approach to the transmission of time information is essential. Several strategies can be adopted for legacy interfaces:

- The transmitting device in each direction provides clock information to the receiver (codirectional). The clock can be integrated to the transmit signal or a separate signal.
- One end (the user or more often the network) provides the clocks for both directions of data transmission (contra-directional). The clock can be integrated to the transmit signal or a separate signal. At the slave end, the transmitter is synchronized by the received clock.
- Both user and network receive independently their clock from a common source.

Protection to TDM Network Synchronization Anomalies

Digital multiplexed communications require the synchronization of receiver clocks for correctly demultiplexing the TDM frame. The clock signal is propagated across the network through the digital signal path. The receiver compares its received clock signal with its own local clock and as long as the quality of the received signal is not below a certain threshold, the received signal is used by the receiver. A communication node with multiple network connections receives several clock signals which are employed according to a preestablished ranking: if the used clock is degraded or disrupted, the next ranking clock is employed, the lowest rank being the equipment's own free running clock. SDH networks operate with a powerful clock information exchange protocol which normally allows the automatic restoration mechanism to operate with appropriate clocking.

© Springer International Publishing Switzerland 2017

C. Samitier (ed.), *Utility Communication Networks and Services*,
CIGRE Green Books, DOI 10.1007/978-3-319-40283-3_15

In the event of a synchronization anomaly which may arise due to human error, equipment malfunctioning, inadequate synchronization plan or certain topological fault situations, the received data may be demultiplexed with an inappropriate (desynchronized) clock at the same "nominal" frequency. Despite a buffer size of one frame (125 μs), the slight difference in frequency (few parts per million) will periodically cause a buffer underflow or overflow, generating a "synchronization slip" (i.e., repetition or loss of one frame). The time period shall depend on the difference in frequency as illustrated below:

- Consider two 2048 kbps clocks at ±5 ppm used for filling up and removing data from a buffer of 125 μs (256 bits at 2048 kps operation).
- The difference in fill and remove frequency is therefore
  2048 kbps × 10 ppm = 20.48 bps.
  Every second the level of buffer moves up or down by 20.48 bits.

- A Buffer Overflow or Underflow shall occur every (256/20.48) = 12.5 s.

Synchronization slips are unnoticed in voice or any other analog services (one errored sample every few seconds) and can pass unnoticed in most data exchange applications due to packet retransmission. Anomalies in synchronization plan (especially under network fault conditions) can therefore exist for a long time in the digital multiplexed network. The impact on Protection communications can, however, be fatal. Any cyclic anomaly in a digital multiplexed communication service must draw the attention to synchronization problems.

Time Distribution—SNTP and IEEE1588v2 (PTP)
Ethernet interface at the Service Access Point allows the exchange of time synchronization information between the user and the network. A packet-switched transport network not being a synchronous communication system like SDH with constant time delay, but a store-and-forward queuing system, the time synchronization is not just required between the user and the network, but between all the devices in the application network at the two sides of the WAN connection. Distributed control and protection applications require a common time reference between intelligent devices dispersed across the network to establish the order of events and to allow coordinated actions when particular grid conditions are recognized. The relative precision of the time reference among devices ranges from 1 ms to below 1 μs depending on applications.

The common time reference can be exchanged implicitly by the user application (e.g., Current Differential Protection), distributed independently from the communication network (e.g., individual GPS clocks or local 1PPS wiring), or increasingly distributed through the local or wide area network through standard time distribution protocols.

In this latter case, a specific IP-based time distribution application is implemented between a time server and its clients. Currently, the Simple Network Timing Protocol (SNTP or NTP) assures an accuracy of 1 ms but entirely depends

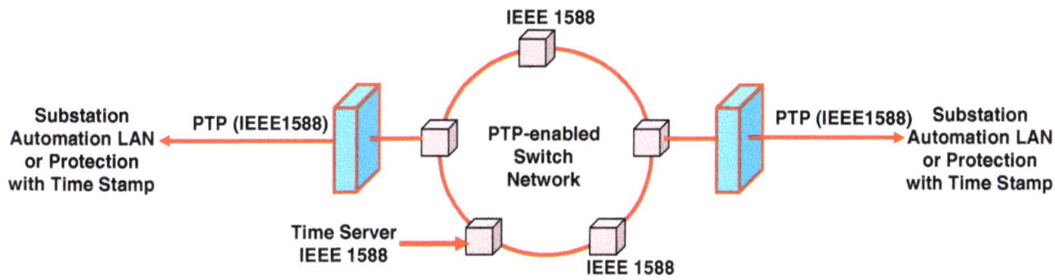

Fig. 15.1 IEEE 1588v2 Precision time protocol (PTP) across the packet switched network

upon delay variations between the client and the time server making it usable essentially inside a star-structured local network (Substation LAN). There is a strong trend to move to IEEE 1588v2 Precision Timing Protocol (PTP), which provides a higher precision of 1 μs (i.e., equivalent to GPS) necessary for more severe automation applications. PTP employs multiple master clocks and each switching node of the network performs path delay calculations. Precision Time Protocol delay calculation mechanism considers that delay variation is only at node level (i.e., queuing files). When multiple switching nodes exist between a master clock and a client, it is necessary that every switch support IEEE 1588v2 synchronization feature. More generally, if 1588v2 is to be supported network-wide then every node of the packet switched communication network that can introduce variable packet delay must be IEEE 1588-enabled or it shall introduce inaccuracy into the timing information (Fig. 15.1).

Circuit and Packet Conversions at the Service Access Point

16

16.1 Packet Over TDM

Transporting Ethernet packets in a time domain multiplexed channel across SDH/SONET communication networks gives a fully dedicated communication channel with high predictability comparable to a dedicated fiber or wavelength, while providing the advantage of a shared transport network infrastructure with its inherent path redundancy and fault tolerance.

Many power utilities use at present SDH networks for transporting operational traffic as E1 or sub-E1 digital multiplexed channels. Migration of these services to packet-based Ethernet can be performed in the first step and with minimal transformation through Ethernet over SDH (EoS).

Despite its lack of scalability, needed mainly at the core of the transport network, Ethernet over SDH provides a very adequate solution in terms of ease of deployment, service migration, field maintenance, and central management for substation access and aggregation layer communications.

Ethernet over SDH (EoS) is of a set of industry standards developed for more optimized mapping and control of Ethernet traffic over SDH. It provides mechanisms for sharing bandwidth with improved granularity and guaranteed Quality of Service. These standards, replacing previous proprietary mapping schemes, provide interoperability and flexibility. The most important ones for Ethernet encapsulation along with TDM transport over SDH networks are listed below:

- Generic Framing Procedure (GFP), ITU-T G.7041, provides a generic mechanism to map variable length Ethernet data streams over SDH synchronous payload. GFP assures that SDH equipment from different vendors transport Ethernet traffic in the same manner.
- Virtual Concatenation (VCAT), ITU-T G.707 allows SDH channels to be multiplexed together in arbitrary arrangements and hence to create pipes at any rate.

© Springer International Publishing Switzerland 2017

C. Samitier (ed.), *Utility Communication Networks and Services*,
CIGRE Green Books, DOI 10.1007/978-3-319-40283-3_16

- Link Capacity Adjustment Scheme (LCAS) is a standard for adjusting dynamically the amount of bandwidth allocated to an Ethernet service.

The simple replacement of an old generation end-of-life SDH system by a new one with enhanced data capabilities responds in many cases to the limited packet-switching requirements in electrical substation networks (e.g., migration of existing SCADA and voice applications into IP/Ethernet in a standard vendor-independent manner).

A very simple and low-cost solution for adding Ethernet into the substation would therefore be the deployment of newer SDH equipment with a service aggregation Ethernet switch separating VLANs for different substation applications.

SDH is indeed no longer the technology of choice for mainstream "operator" telecom deployment, but the lower end constituents of the hierarchy (<STM-16) still remain for some time a valid option for time-sensitive private networks or for their substation access component.

16.2 Circuit Emulation Over Packet

Diagonally opposite to Packet over TDM discussed in the previous section, one can find TDM over Packets which is the possibility of transporting some limited amount of circuit-type traffic over a packet-switched transport network or substation access interface. In electrical power utility telecom networks, TDM over Packet provides a way to transport legacy applications, originally designed for TDM transport, over a network which is transformed into packet-switching. This is a race of packet-transformation between Application Networks and the Transport Network.

The most critical example for TDM over Packets is the case of protection relay over a packet network. Despite the recent introduction of IEC61850, the protection relays in operation today are designed for circuit-mode communications and will remain on the network still for many years. If the only available transport network is packet-based, then the conversion can be performed at different levels as shown in Fig. 16.1:

- Protection can be connected to a TDM aggregation device (Access Multiplexer) together with other legacy circuits and transformed into a 2 Mbps E1 interface. The transport network undertakes the delivery of an E1 Service Access Point which it transports over a packet-switched network (PSN) incorporating TDM circuit interfaces in its User-to-Network interface (UNI). The transport network SLA is that of the E1 with appropriate qualities (delay, loss, etc.).
- The aggregation device, a multi-service access platform, performs the necessary transformations to present a packet interface (e.g., Ethernet) requiring an Ethernet Tunnel (Ethernet Line Service) from the transport network.

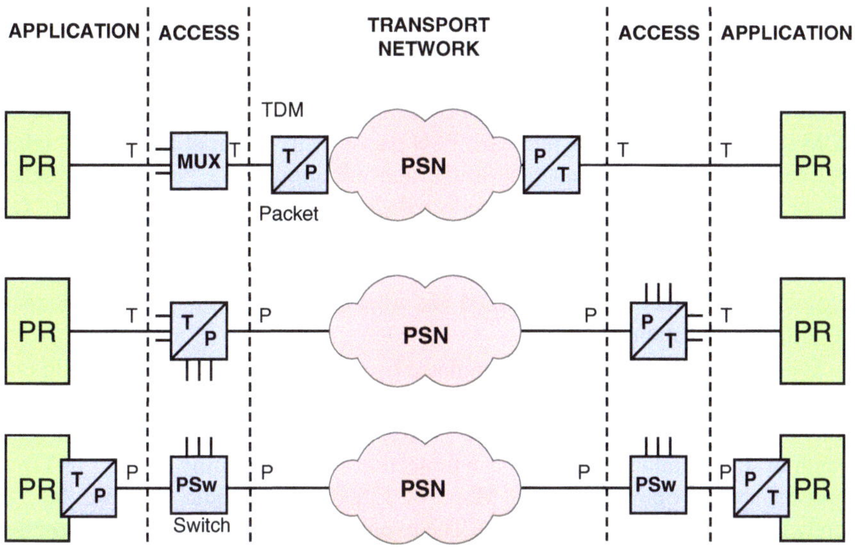

Fig. 16.1 Transporting legacy protection over a packet-switched network (T = TDM, P = Packet)

- An Ethernet encapsulation device can be inserted into the protection communication port at the application side presenting an Ethernet interface to service aggregation (Ethernet Aggregation Switch) for presenting an aggregated Ethernet stream at the Service Access Point.

TDM Pseudowire Emulation Principles

Time division multiplexing can be emulated across a packet-switched network through encapsulation of synchronous data into non synchronous packets at the emitting side, and buffering of asynchronous data to reconstitute the synchronous TDM signal at the receiving side (de-jittering buffer). TDM Pseudowire Emulation (PWE) mechanisms emulate the attributes of a TDM service such as an E1, T1 or a fractional $n \times 64$ service. These are described in a number of IETF RFCs, the most appropriate for time-sensitive communications being RFC4553—Structure-Agnostic Time Division Multiplexing over Packet (SAToP). This is currently implemented into various packet-based multi-service access and transport nodes (e.g., Ethernet Switch, MPLS Router, etc.).

The quality of the resulting TDM interface depends on "how well the TDM channel is being emulated" (i.e., channel performance) and consequently on adequate parameter adjustments, traffic engineering, buffer dimensioning, and bandwidth resource allocation.

The main constraint discussed here is the time behavior of the channel, although channel availability and behavior upon failure can also be decisive especially when one has to rely upon an external provider's Service Level Agreement (SLA).

Packet Encapsulation

An important factor in PWE for time-sensitive legacy applications is the Packet Encapsulation Delay (PED). This is the time delay introduced by buffering of data before transmitting them as Ethernet Packets (e.g., minimum 64 Bytes). The more TDM data samples we pack into each PSN packet (e.g., Ethernet frame) the higher will be the PED (due to wait time in the transmission buffer), but also the higher will be the bandwidth efficiency (due to reduced idle capacity in each packet). On the other hand, we can pack very little TDM data (e.g., only one sample) into each packet so that it can be expedited faster, but in this case, the remaining capacity (e.g., of the Ethernet frame) is wasted and a lot of bandwidth is required for each TDM communication channel.

Typically, a 64 kbps channel is emulated by sending 2 bytes of TDM data in one Ethernet frame, which results in an Ethernet frame rate of 4 kHz. The resulting Ethernet frames will have a length of 64 bytes, which corresponds to the minimal length of such a frame, meaning that a protection system requiring 64 kbps of TDM capacity shall require at least 2 Mbps of Ethernet bandwidth in PWE (i.e., a bandwidth efficiency of 2/64 or 3 %). The remaining bytes may be used for sending auxiliary information (e.g., management and monitoring) which do not necessarily exist in any evolved manner in legacy application systems. Sending more bytes of primary information results in increased transfer delay due to transmit side buffering. It can be seen from Fig. 16.2 that the packet encapsulation is feasible when high bandwidth is available and low efficiency can be afforded.

Clock Recovery and Jitter Buffer Delay

Packet-switched data networks, not being initially designed to transport TDM signals, do not have any inherent clock distribution mechanism and the clock signal inherent to TDM transmission is lost in the encapsulation process. When transporting TDM, the receiver must therefore reconstruct the transmitter clock by phase-locking a clock to the fill level of a Jitter Buffer (receiver data buffer) as shown in Fig. 16.3. Threshold detection in the "de-jittering" reception buffer is employed to construct a synchronization signal. The buffer size used for this purpose must be such that no underflow causing synchronization slips can occur. The buffer size is therefore dependent upon the maximum delay variation of the communication channel which is to be absorbed through buffering. The jitter buffer, however, generates absolute delay (Jitter Buffer Delay or JBD). No need to say that the larger the buffer, the more it introduces absolute delay. In order to maintain a

Packetization (bytes/packet)	Bandwidth (kbps)	PED (ms) for 64 kbps TDM signal
2	2048	0,25
4	1024	0,5
8	512	1
16	256	2

Fig. 16.2 Packet encapsulation

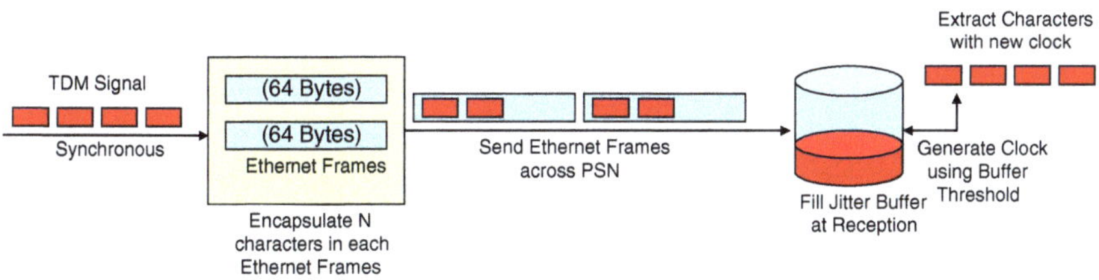

Fig. 16.3 TDM pseudowire emulation

low buffer size and hence the JBD, it is essential that the delay variation of the packet-switched network be kept as low as possible. A dedicated point-to-point Ethernet connection over fiber or an Ethernet over SDH (EoS) typically provides a very low delay variation and therefore a fairly low JBD. On the other hand, a large packet network with many concurrent users may give considerable delay variation and hence a large JBD.

Moreover, it should be noted that clock recovery through Jitter Buffer threshold detection may be vulnerable to packet loss in the transport network although different techniques have been employed to overcome this issue.

In a private dedicated packet network it is also possible to associate a synchronization clock to each packet-switching node and hence to render the Ethernet packet network fully synchronous. This technique, known as Synchronous Ethernet (SyncE), is further discussed in Part 4 on network implementation. With a jitter buffer size of 8 Ethernet frames (64 bytes each), and an Ethernet connection at 2048 kbps, the JBD shall be 2 ms and it shall be capable of absorbing a maximum delay variation of up to ±1 ms.

To conclude on circuit emulation over a packet network, the total time delay (TD) is given by

$$TD = PED + JBD + T_P + ND$$

TD	Total delay	PD	Processing Delay = typically 10 μs
PED	Packet encapsulation delay	SFD	Store and Forward Delay = FS/BR
JBD	Jitter buffer delay	FS	Frame size (FS_{max} = 1500 bytes)
T_P	Propagation time	BR	Link bit rate
ND	Network delay (QD + PD + SFD) × N	N	Number of PSN nodes
QD	Queuing Delay = SFD_{max} × Network load		

Meeting a tight level of Time Delay and Delay Symmetry, as for Protection communications, requires careful adjustments and dimensioning of parameters across the packet-switched network.

Modeling the Service Delivery Process 17

Service delivery is not just about architecture and interface coordination, but also a process enabling the organization to provide satisfactory communication services to the users. Formal process frameworks and models such as ITIL (Information Technology Infrastructure Library) are currently employed in many power utilities IT organizations implying familiarity, practice and trained staff. ITIL defines a framework for internal processes necessary for running an IT infrastructure. It classifies tasks and processes into groups and specifies interactions between them. ITIL is relevant for management of the IT services, platforms, and systems in any enterprise context including that of the telecom service delivery organizations.

Business processes allowing service delivery by a Telecom Service Provider, on the other hand, have been specified through a Telecom Operations model called eTOM produced as part of a solution framework NGOSS (Next Generation Operation Support Systems), devised by TeleManagement Forum.

These frameworks have been reviewed by CIGRE D2 in order to build a reduced structural framework named "Utilities' Telecom Operations Map" (uTOM), applicable to the domain of Utility telecommunications and comprising a well-defined set of processes to formalize the associated roles and the interactions. Here, although we do not intend to reproduce a full account of this work, a short introduction of the uTOM model is necessary in order to use it for a structured discussion of service delivery process in this section as well as implementation processes in part 4 and operation and maintenance of network infrastructures in part 5 further in the book. Readers may refer to the initial document for a more detailed analysis.

The uTOM model, shown in Fig. 17.1, distinguishes three blocks of processes and activities

© Springer International Publishing Switzerland 2017

C. Samitier (ed.), *Utility Communication Networks and Services*,
CIGRE Green Books, DOI 10.1007/978-3-319-40283-3_17

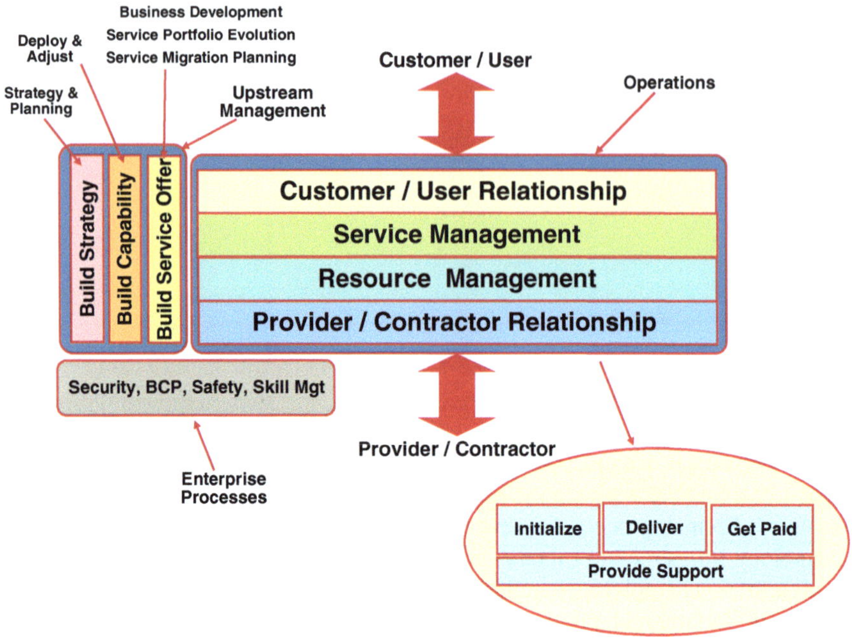

Fig. 17.1 Utility telecom operations map uTOM

1. Upstream management comprises business planning and development, specification of deliverable services through a service catalog, and capability building, through network, tools, processes, contracts and partnerships, for delivering the services. This group of processes delivers an organization with staff, processes, assets, and service contracts allowing the delivery of telecom services within the EPU. It also comprises the migration process for the transformation of delivery scheme, service, network or organization to meet new requirements or to improve the quality or the cost of the delivered service. Upstream management is further discussed in the following part on network implementation.
2. Operations consist in current, day-to-day actions for delivering telecom services and maintaining the network and its related resources to provide user connections meeting users' service requirements. This latter group of processes comprises the establishment, operation, monitoring, and maintenance of logical and physical connections over the deployed network.
3. Transverse enterprise processes such as skill management, IT security, and business continuity, existing at enterprise level, become particularly critical when applied to the telecom network operational organization and hence constitute a third block not specific to the telecom service.

The focus of the present section is the current operations which is further discussed in part 5 of the book. The operations process interacts with the outside world through two external interfaces:

- service users to whom the organization delivers communication services
- service providers and contractors from whom the organization procures all or part of its resources (connectivity service, network infrastructure, tools, and workforce) in a recurrent manner (building the network is indeed not a recurrent activity).

Operations Management process block is composed of four transversals as shown in Fig. 17.1:

- Customer/User Relation Management comprises all processes governing information exchange with internal or external telecom customers for delivering communication services with preestablished level of quality (standard SLA) and hence obtaining user satisfaction. This is the customer view of the Telecom Management.
- Communication Service Management covers all processes required for the delivery and support of communication services to customers over company-owned resources and/or externally procured services. In particular, it should be noted that even in the case of a fully procured service scheme, a "thin" telecom service layer is still required to ensure the adequacy between user requirements and provider service obligations (and possible association of multiple procured services to meet the requirements).
- Network Resource Management covers the management processes relative to network infrastructure and associated tools and applications which are employed for delivering the service. This horizontal corresponds to the network view of the management process.
- External Providers and Contractors Management covers all processes concerning relationships to contractors, partners, and suppliers whose services enable the organization to provide telecom services or to maintain the network infrastructure used for telecom service delivery. The externally procured or contracted services can be technical support, field maintenance, dark fiber, bulk transmission capacity, or leased application-level connectivity.

Operations management processes can also be related to a lifecycle track "Initialize, Deliver, Provide Support and Get Paid" (Fig. 17.1) corresponding to the eTOM verticals:

- **Fulfillment** (**Initialize**)—Setting up and initialization of different schemes (users, services, channels, tools, etc.) based upon preestablished agreements and rules.

	Fulfillment (Setting up)	**Assurance (Delivery) *Running***	**Support (Enquiry Supervision & Maintenance)**	**Accounting (Metering, Cost Repartition, Invoicing & Settlement)**
User/Customer	User Order Handling User Change Handling	User Problem Handling User SLA Management	Service Enquiry Desk (User Technical Support)	Customer/User Entity Invoicing & Settlement
Communication Service	Service Configuration & Activation	Incident Management Service Quality Mgt	Service Inventory/Config. Mgt Service Change Management	Service Policing & Usage Metering
Network Infrastructure & Resources	Bandwidth & Capacity Provisioning	Network Problem Mgt Network Perf. Management Disaster Management	Net. Configuration Mgt Network Change Mgt Fault Management Network Maintenance Asset Lifecycle & Spare Mgt Management Tools Support	Estimate running cost of network infrastructure (+ tools)
External Providers/ Contractors	P/C SLA Setting & Adjustment	P/C Performance Mgt (SLA Monitoring) P/C Problem Reporting (Supplier Relationship Mgt.)	P/C Support Management (Site Access Permits, Safety & Certification)	P/C Cost Assessment & Invoice Settlements

Fig. 17.2 "Current operations" process for utility telecom service delivery and provision (Telecommunication Service Provisioning and Delivery in the Electrical Power Utility 2011)

- **Assurance (Deliver)**—Running the schemes set up previously in order to deliver the communication services.
- **Support (Provide Support)**—Providing a response to different enquiries and requests for service including maintenance services.
- **Accounting (Get Paid)**—Determination and recovery of costs related to the delivery of telecom services to different Utility-internal or external service users according to rules which are set by the upstream management processes, and monitoring of the usage of communication resources according to contracted conditions and recovering of network revenues to settle the operational expenditure.

Figure 17.2 presents a structured list of processes identified across the domain of operations management in the power utility telecom context.

Referring to the typical utility telecom organization scheme presented in part 5, the following mapping can be performed between the process model and the organization model:

- Network planning and transformation team covers the upstream management perimeter. In particular, changes of the network infrastructure, communication services, and management tools are planned and managed through this organizational block.
- Customer/user relations management is the core activity of the service management and user relations organization (or person). This entity is the single point of contact for user entities in order to request new connections and to signal service anomalies. It is also responsible for notifying user entities of any service interruptions and to produce quality of service dashboards.

- Communication service and network resource management are the tasks performed by the network supervision and maintenance and support entities. These entities maintain the technical quality of the service across the network infrastructure.
- Finally, external providers and contractors are managed by maintenance and support and/or service management depending upon the type of the procured service.

Managing the Delivered Communication Service 18

In this section, we have compiled all actions performed in the day-to-day operation of the network without dealing with the infrastructure itself. It covers the two top layers of utility telecom operations map in Fig. 17.1 though the tasks relating to creation and modification of services is integrated into the section on change management and configuration in part 5. In addition, service management can also cover the formal relationships to service providers from whom the utility provisions all or part of its services. This can be performed through the same team or person managing service users or a separate team.

The compilation of tasks can correspond to the role of one person (or a small team depending on the scale of operations) which we shall call the service manager and will be necessary whether the infrastructure is owned and operated by the utility or contracted to a third party. Some major constituents of this compilation are described below.

Service Desk

Service Desk is the interface point for user relations. In the majority of utilities it is materialized through telephone and/or e-mail, although web-based automated service desk may also be employed in larger organizations with merged telecom and IT activities.

Service desk is the point of entry for user-detected anomalies whose extent is widely variable across the surveyed utilities. The trouble ticket for user-detected incidents is opened and introduced into the system by the telecom service providing staff and not by the service user. The service desk must, however, comprehend the user problem and follow its resolution in order to notify the enquiring user and to generate the adequate report.

The service desk tracking of user issues is a valuable tool for monitoring user satisfaction, providing statistical data for the analysis of services, resources and the evolutions of user expectations feeding both continuous improvement and service migration processes.

© Springer International Publishing Switzerland 2017

C. Samitier (ed.), *Utility Communication Networks and Services*,
CIGRE Green Books, DOI 10.1007/978-3-319-40283-3_18

Quality and SLA Monitoring, Notification and User Dashboard
Monitoring the quality of the communication service delivered to utility users comprises two distinct components

- Monitoring through User Applications—This is a customer feedback on the perceived quality (e.g., power system control center) which can be used by the service provider as a measure of the technical performance of the system. A continuously communicating application such as SCADA can deliver monitoring data every few minutes and can therefore be used to detect performance anomalies due to hidden faults and configuration errors. It can also be used to estimate user service availability and SLA fulfillment.
- Monitoring through Impact Analysis—The service provider can determine the service impacts of detected network faults through its monitoring of the infrastructure. Service impact analysis capability integrated into some management platforms, allows the backward usage of root cause analysis translating the unavailability of a resource (e.g., cable segment, telecom equipment, physical, or logical link) into a potential outage or degradation of a communication service delivered to a user. In many cases, the user may be unaware of this outage (e.g., for discontinuous usage of the service) or degradation (e.g., loss of a backup route). Service provider's monitoring of service availability can be used to check and prove SLA fulfillment possibly against user's application-level monitoring of service availability.

User notification is the mechanism used for alerting a service user about service impacts detected by the provider. It can be automatic (upon service impact detection) or manual. In both cases, phone call, SMS, and mail can be adequate means of notification.

In addition to service availability and outage measurements, the service provider must monitor the metrics constituting the SLA for any particular type of communication service. Depending on service requirements, this may include

- throughput,
- time delay (latency) and delay variation,
- information loss and error,
- service restoration time upon fault, and outage,
- problem resolution time upon incident reporting, etc.

The results can be made available to the service user entity as a **User Dashboard** which allows quality reporting to service users with user-relevant information (user view of the communication network).

Continuous monitoring of SLA metrics further allows prompt detection of contractual SLA violations allowing appropriate priority changes in the teams in order to avoid major application-level incidents and consequent contingencies (e.g., for critical applications such as protection relay communications). It should be noted that for a majority of surveyed utilities, at the time of preparation of this book,

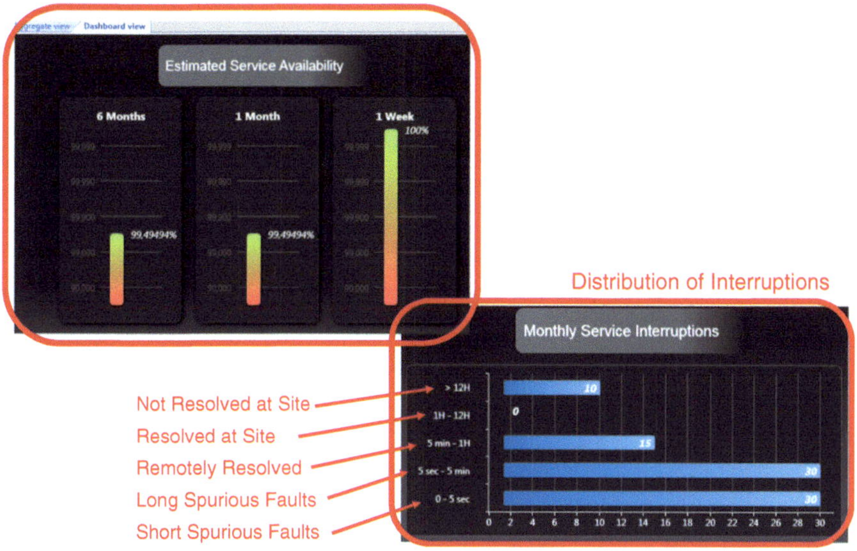

Fig. 18.1 Service availability and outage statistics as part of a Service Dashboard (Courtesy of GE Grid)

formal SLA and contractual obligations only concern the relationship to external providers (e.g., O&M contractor) and to external service users (Fig. 18.1).

Service Reports, Statistics, and Trends are at present mostly generated manually using data and reports from various management systems and using office tools such as MS Excel. An integrated management platform capable of service monitoring generally includes statistics generation and trend analysis capabilities reducing substantially the effort required for building internal or contractual service reports.

Usage Measurement and Invoicing

Although in the great majority of power utilities operational services are not directly paid by the user entities, with the sharp growth of communicating applications, it is increasingly important to have a precise estimation of the bandwidth and the services allocated to each user entity in order to regulate network usage and to partition the cost of maintaining, upgrading, and extending the network and its services among the interested parties across the power system organization. The bandwidth usage estimates can be produced by the infrastructure management and network configuration's bandwidth policing and inventory.

Whatever be the situation in terms of user invoicing of services, the O&M organization needs to assess costs and supervise the quality of service of its service providers and contractors including the measurements of the utility's usage of these external services. This includes procured communication services, procured usage of infrastructures, and procured maintenance and field services.

Meeting Service Quality at a Packet-Switched Access Point 19

Meeting contractual service qualities at a time multiplexed service access point is relatively straightforward: the performance is fully predictable in normal operation and constant once met at the hand-over of the connection to the user. With packet-switched services, the situation is quite different. The technical performance that can be expected depends upon the precautions taken in the resource sharing scheme and the overall bandwidth of the resource. Other issues such as service restoration mechanism play also an important role in the overall expected qualities of the delivered service.

This present section gives only an overview of some of these issues, while a more detailed discussion is beyond the scope of the present book and can be found in many network design handbooks. Some fundamental principles of packet switched and time multiplexed communication technologies and corresponding network architectures are provided as a reminder in part 4 of the present book.

Quality of Service (QoS) and Network Performance objectives

Performance issues for packet networks have been extensively analyzed and documented in the public telecommunication domain with application imperatives which are somehow different from our present operational communications context.

The ITU-T Recommendation Y.1541 titled "Network Performance objectives for IP-based services" defines classes of network Quality of Service (QoS) as the basis for agreements between end users and network providers. The ITU-T document then defines network performance parameters, a reference network model and specifies network performance objectives for different QoS classes.

For protection and control communications between substations, IEC61850 proceeds more or less in the same manner, defining application models and functional interfaces, and then communication requirements through performance classes (refer to IEC61850-90-1 Sect. 6.3 and IEC61850-5 Sect. 13.4). However, in the present state of IEC61850 documents, the physical network reference model is not sufficiently precise, QoS classes not rigorously specified and there is a permanent confusion between user's QoS and network's performance.

© Springer International Publishing Switzerland 2017

C. Samitier (ed.), *Utility Communication Networks and Services*,
CIGRE Green Books, DOI 10.1007/978-3-319-40283-3_19

In the present section, methodology and parameters are taken from ITU-T Y.1541 (with different protection relaying communication services instead of IP voice, IPTV, and file transfer services. Moreover, the exclusively IP referencing of performance parameter in ITU-T document is removed in favor of a more general packet communication, considering that in time-sensitive applications such as protection relaying, we are more often concerned with Layer 2 Ethernet rather than Layer 3 IP-based services.

Quality of Service is the ability to guarantee a certain level of performance to a user data flow and hence meeting the network user's requirements. It is related to the network technical performance (e.g., switching speed, network throughput, etc.) but is also related to the mechanisms which are used for distinguishing different traffic flows in a multi-service environment and assigning priorities to each of them. In the IP networking context, the term QoS very often is mainly used to define these service differentiation and priority assignment mechanisms.

In defining QoS objectives, one important ITU-T Y.1541 concept is the **Hypothetical Reference Path** (**HRP**) defined as a typical end-to-end path taken by the user data flow from the source host User-to-Network Interface (UNI) to the destination host UNI. This UNI-to-UNI path is used for setting performance objectives.

In the case of utility operational communications over a packet-switched network, the HRP does not include the crossing of multiple networks but different cases must still be distinguished depending upon architectural options as illustrated for protection relays in Fig. 19.1:

- Ethernet-interfaced application connecting to its remote counterpart through a dedicated fiber, wavelength or a point-to-point unshared Ethernet connection across one hop of a TDM network (e.g., Ethernet over SDH).

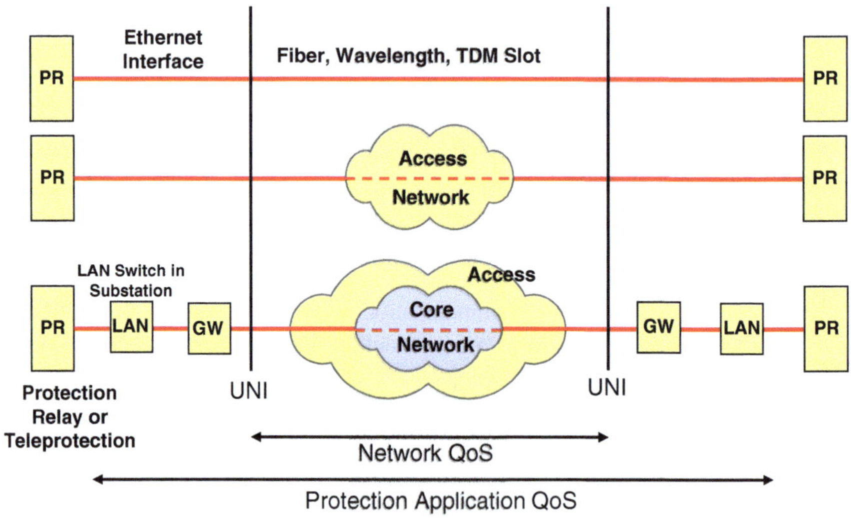

Fig. 19.1 Protection communications hypothetical reference path models

- Ethernet connection including intermediate switching nodes across the substation aggregation and access subnetwork.
- Ethernet connection including substation-level LANs at each end (IEC61850-based architecture) for accessing the application device (e.g., protection relay).
- Ethernet connection that requires the crossing of a core network in addition to the substation aggregation networks at each end (e.g., data network shared between operational and non-operations users, or Ethernet service provisioned through a public data network).

The main performance parameters defining objectives in a packet network are as follows:

- **Packet Transfer Delay (PTD) or Transfer Time**—The mean transfer delay is the sum of the mean delays contributed by network sections (LAN section at each side, the access or distribution network and the core network if applicable). At very slow serialization rates, packet insertion times can predominate the network transfer time.
- **Packet Delay Variation (PDV)**—Packet Delay Variation can be defined as the difference between maximum and minimum PTDs in a given time interval. Relating delay variation to individual network section values is difficult due to the probabilistic and nonadditive nature. Supposing Gaussian (Normal) distribution for the PTD, delay variation is related to the variance and can therefore be estimated for a multi-segment network. The objective is often based on the difference between a maximum PTD value that is not exceeded with a given probability figure (e.g. 10^{-4}) and an equally probabilistic minimum value. Another delay related parameter in packet networks that could be added is delay asymmetry already discussed in part 2, although this is not specified as an objective in ITU-T Y.1541. Admitting that the PTD is a random variable with a mean value (PTD) and a probabilistic variation (PDV), we accept that there will be a probabilistic asymmetry between go- and return-directions' transfer times. Delay variation having identical statistics for the go- and return-directions then equal size reception buffering shall be implemented at both ends which are, however, not synchronized.
- **Packet Loss Ratio (PLR)**—Packet loss ratio is the upper bound on packet loss probability. If multiple network sections are present, then this is estimated through the probability of no loss which is the product of probabilities of no packet loss in each cascaded section. For this objective, it is necessary to specify a maximum waiting time to declare a packet as lost.
- **Packet Error Ratio (PER)**—This is estimated by inverting the probability of error-free transfer which is as in the previous case, the product of probabilities of error-free transfer in each network section. PER naturally depends upon packet size as well as the error-generating aptitude of the communication channel. The detection and correction mechanisms in the network transform this PER into a packet loss (errors which are detected but not recovered).

- **Packet Route Control (PRC)**—Packet Delay Variation is not the only source of Delay Asymmetry. Nonidentical path, in particular upon route change, (not presented in the HRP) also cause delay asymmetry. It is therefore reasonable to put the extent of Routing Control as a performance parameter used to define different QoS objectives for protection applications.
- **Packet Route Restore Time (PRR)**—For service availability, ITU-T stipulates that "further study is needed since design options are rapidly changing." Statistical availability achieved through robustness of components, reliability of subsystems and of transmission media is a constant for all automation applications in the substation. It is therefore not a network performance for specifying different classes of QoS. Service restoration time, on the other hand, is a determining factor for selecting packet technologies and network mechanisms guaranteeing the continuity of operational communication services.

Network Resource Sharing, Traffic Queuing, and Priority Assignment

Packet networks are fundamentally based upon the idea of resource sharing to provide a more efficient usage of network resources such as transmission bandwidth. There exist two trivial ways for meeting the performance requirements of critical services in this context

- **Dedicated network resources** for every critical service (not shared with other applications). This includes dedicated fibers, wavelengths or time slots. In other words, takes out the critical services from the resource sharing scheme provided by packet switching and hence reducing the bandwidth efficiency. This is the simplest and most secure way of assuring QoS, already extensively used in protection applications.
- **Over-provisioning** of network resources provides traffic queuing but no packets in the queues. This does not provide a guarantee of meeting QoS requirements for any class of traffic but reduces the probability of delay due to queuing.

In addition to these trivial methods, other QoS control mechanisms can be employed such as:

- **Priority Assignment**—used in native Ethernet but also in IP-MPLS with no Traffic Engineering. Different Classes of Service are distinguished (IEEE 802.1Q) using a priority indication field. Priority assignment in switched Ethernet means that different priority traffic flows join different queues. One should, however, note that a low priority packet whose service is in progress is not dropped when a high-priority packet arrives in the queue. It may therefore have to wait for a "maximum-length" lower priority packet to be served before profiting from its priority. Similarly, if multiple traffic flows are assigned high priority, then they will be in the same queue and the High-priority packet may wait many queued packets. Some Ethernet switches, however, allow, through supplementary queues, to distinguish between "transit" and "add" traffic so that, for traffic at a particular priority level, the "through" traffic (line to line) has priority over the "add" traffic (drop to line). This mitigates to some extent the delay accumulations over multi-hop paths.

- **Resource Reservation (RSVP)**—Host or network node requests network resources dynamically for a particular communication. RSVP is an upper layer protocol (transport/application) employed in some IP networks and for the distributed control plane in some MPLS networks (MPLS-TE). It supports end-to-end QoS control and is complex to implement.
- **Traffic Engineering** mechanism provides Peak Information Rate (PIR) and Committed Information Rate (CIR) which can be set for different traffic flows. This allows adjusting constant traffic for some critical flows (CIR = PIR), a minimum throughput for some others (CIR < PIR) and best effort for some other flows (no CIR). In MPLS-TP, the centralized control plane sets statically the QoS control parameters for different traffic flows in a coordinated manner. This mechanism provides quasi-deterministic transport service for the most critical services.

Differentiated services (DiffServ) quality control used in IP networks determines a "Per Hop Behaviour" according to a Type of Service (ToS) field qualifying the data flow as EF (Expedited Forwarding) for minimal latency and jitter for time-sensitive IP traffic such as IP voice and BE (Best Effort) for non-time-sensitive web-traffic. DiffServ does not support end-to-end QoS control but constitutes a contractual element for service delivery.

Packet Route Restoration

The existing predominant transport technology widely used for operational communication being SDH/SONET, the corresponding standard-specified restore time figure of 50–60 ms is generally used as a reference value for the restoration time in any other new technology in this network.

IEC61850-90-1 (paragraph 7.5), however, stipulates that "Unless dual-port IEC 61850 IEDs are used with physically separate paths, the Ethernet network should recover (restore traffic) from a fiber failure within 10 ms." This statement has large implications on transport network architectures and technologies, favoring partitioned rather than end-to-end packet route restorations. Main restore mechanisms are presented hereafter and summarized in Fig. 19.2:

- **SDH protection mechanisms**—Packet service on an underlying SDH layer used for protection communications can use route restoration based on SDH inherent ring and linear protection mechanisms. The achievable route restore time is in reality significantly shorter than the standard specified 50 ms (typically 15 ms).
- **RSTP (Rapid Spanning Tree Protocol)**—IEEE 802.1w allows recovery time of around 5 ms per switch in the network. No more than 7–10 switches are to be included in an RSTP automatic recovery subnetwork, favoring a hierarchically partitioned Ethernet network as described in Part 4.

- **MPLS-TP Linear and Ring Protection Switching**—MPLS-TP supports traditional transport networks protection mechanisms switching in less than 50 ms. Moreover, it uses the same path for both go- and return-directions both in normal and alternate paths and is therefore suitable for symmetry-sensitive protection relay systems.
- **IP/MPLS**—These networks use the Fast Reroute (FRR) mechanism using the traffic engineering properties of MPLS. Backup paths are determined for each MPLS route at each "point of local repair" where traffic flow can be redirected in the event of local detection of anomaly at the following node.
- **IP Routing Resilience**—Resilience inherent to IP routing algorithms such as OSPF provide great flexibility and the capability of being used in very large networks, but completely unsuitable for applications requiring fast restore and route control.
- **HSR (High-availability Seamless Redundancy)** and **PRP (Parallel Redundancy Protocol)** are considered at present as LAN restoration mechanisms (could be applied in the wide area).

Synchronization and Clock Distribution

Maintaining a common time reference between IEDs is a strong requirement in many substation automation applications in order to establish the order of events, to correlate events, and to allow coordinated actions when particular grid conditions are recognized.

The level of relative time precision required among cooperating elements is variable and depends upon applications. Moreover, the cooperating elements may be located in the same substation (intra-substation automation system) or located across the grid (inter-substation or substation-to-control platform). SCADA applications, for example, generally require an accuracy of 1 ms to deliver chronological sequence of events with adequate granularity. Protection and control applications sampling the 50–60 Hz power waveform would generate unacceptable measurement errors with a 1 ms timing precision (e.g., at 50 Hz, 50 μs time shift generates a phase error of one degree).

Packet Technology	Service Restoration Time & Mechanisms
Ethernet over Separate Fiber or Separate Wavelength	Linear Protection Switching, <50ms
Ethernet over SDH or OTN	SNCP Ring Protection, <50ms
Shared Ethernet (VLAN)	RSTP (Rapid Spanning Tree Protocol) IEEE 802.1w , > 1 sec
Ethernet over IP/MPLS	Fast Reroute Restoration (FRR), 50ms
MPLS-TP	Linear/Ring Protection Switching Programmed through Management System, 50ms
IP Routing	IP Routing algorithms (OSPF, etc.), >1 sec

Fig. 19.2 Service restoration capability of packet technologies for protection relaying

For time-sensitive applications, time synchronization is necessary at different levels

- Time stamping of event indications from the protection relay and teleprotection signaling to local event recorder and other components of local substation automation.
- Time stamping of analog measurement samples in current differential protection and synchrophasor-based system integrity protections for the decision process at the remote end protection relay or WAPAC.
- Time stamping of commands to be transmitted to remote substation components for protection systems or for remedial action schemes.

Local time stamping of events in the power system has been realized for many years through the distribution of IRIG-B or 1PPS signals via dedicated wiring or through installation of GPS clocks for each device.

Current differential protections have mainly used their own implicit time synchronization scheme based on the assumption of fixed and symmetrical (go-return) communication channel delays, replaced sometimes by GPS clocks.

One major drawback of GPS synchronization, however, is its nonstandard distribution concept. Substations with a large number of GPS-synchronized IEDs end up not only costly but as a "forest of GPS antennas" or alternatively a network of nonstandard GPS antenna splitters.

The proliferation of Ethernet LAN in the substation brought Network Timing Protocol (NTP) in the substations and IEC61850 specified the usage of IETF RFC2030, known as **Simple Network Time Protocol (SNTP)**, for time synchronization with multiple performance classes covering different substation automation time tagging requirements.

SNTP clock distribution is based on the exchange of specific time packets through an SNTP application layer (above UDP/IP) between a time server and its clients (IEDs). The time client controls the interval between two time requests from the server. SNTP defines both unicast and broadcast time distributions but IEC61850 specifies only the usage of the unicast mode.

Considering the fact that SNTP in normal network conditions can at best provide an accuracy of 1 ms, lower level time stamping is sometimes used between time server and clients to meet more severe substation automation requirements. Moreover, since the accuracy of the time distribution scheme depends upon transmission time variation, the number of hops between the time server and the IED must be as small as possible.

IEEE 1588 Precision Time Protocol (PTP) allows 1 μs time accuracy which is equivalent to that presently achieved through GPS and meets in this way the of more severe substation requirements such as IEC 61850-9-2 Process Bus or synchrophasor-based applications. IEEE 1588 can have multiple master clocks, the timeserver initiates the time update in the clients, and the protocol comprises path delay calculation mechanisms. The result is a sub-1 μs accuracy time distribution system over Ethernet networks. The accuracy of IEEE 1588 is equivalent to IRIG-B or PPS without the need for dedicated cabling. However, it should be noted that IEEE 1588 (as SNTP) is an IP-based protocol and its performance relies upon the design and operation of the telecom network and in particular on available bandwidth. In order to meet the microsecond accuracy, all the network devices must support this standard and this may necessitate the addition of specific synchronization hardware. IEEE 1588-2008 (also called IEEE 1588v2) improves accuracy and robustness using a master-slave hierarchy to deliver frequency, phase, and time of day information.

ITU-T Synchronous Ethernet (or SyncE)
ITU-T G.8262 defining Sync-E is an alternative way for ensuring high accuracy time synchronization. Sync-E is not a time distribution protocol like SNTP or IEEE 1588 but a total change of philosophy of packet networks back to the telecom fundamentals of synchronous digital communications and even the primary E1 multiplexed signal. It uses the Ethernet physical layer to pass timing information from node to node in the same way as done in SDH/SONET.

Synchronous Ethernet locks the timing of the Ethernet physical layer much in the same way as SONET/SDH. The physical layer transmitter clock is derived from a high quality frequency reference traceable to a primary reference clock instead of the ± 100 ppm crystal clock of the Ethernet node. The receiver, at the other end of the link, locks to the physical layer clock of the received signal. Then, in TDM fashion, this receiver locks the transmission clocks of its other ports to this frequency reference. In this manner, by feeding one network element in the Ethernet network with a primary reference clock, and employing appropriate timing recovery in the Ethernet physical circuitry, we can set up a frequency-synchronized network (but not necessarily time-synchronized).

Implementing a Synchronous Ethernet solves many of the problems of legacy protection over packet networks and TDM pseudowire emulation described previously, and simplifies accurate time stamping issues of this section. However, it requires that all switches involved in the network be equipped with adequate oscillators and timing recovery mechanisms.

Multi-casting and data flow filtering
Multi-casting is the capability of simultaneous transmission of a message to multiple destinations and is essential to a great number of PSN-based protection applications. GOOSE messages and time-stamped measurements often require to be sent as multicast.

Sending a message to multiple destinations is an implicit property of Ethernet networks which naturally operate through broadcasting of all received frames across the whole network. When Ethernet connectivity service is carried over a point-to-point transport infrastructure (e.g., Ethernet over SDH) or emulated over another network (e.g., over IP/MPLS), then multi-casting mechanisms have to be devised in order to meet such requirements. In the case of “native” Ethernet switching, the problem is not devising a broadcast mechanism, but to restrict traffic flows only to network segments and devices which require access to that particular traffic flow.

The most common means to configure flow restrictions in a switched Ethernet network is implementing VLANs between all IEDs requiring access to a certain data flow. In this way multicast traffic is limited to the VLAN broadcast domain.

Certain Ethernet switches further have the capability of using higher layer protocols (above IP) to manage multicast groups and hence reduce multicast traffic and unnecessary packet processing. IGMP (Internet Group Management Protocol) is an integral part of the IP multicast specification, operating above the network layer to manage multicast group memberships. Network switches supporting "IGMP Snooping" analyze all IGMP packets between hosts connected to the switch and multicast routers in the network. When a switch hears an IGMP report from a host for a given multicast group, the switch adds the host's port number to the multicast list for that group. In substation applications, IGMP allows for multicast data such as GOOSE frames to be filtered and assigned only to those IEDs which request to listen to them.

Integrating Service Delivery for IT and OT Communications

20

A recurrent issue in planning power utilities telecoms is whether communications relating to the grid operation and those relating to enterprise administration should be delivered by the same network.

The enterprise network of the utility is an Integrated Service IP network covering corporate enterprise, contractual and commercial communications in utility offices and information platforms, often with extensive connection beyond the enterprise perimeter (e.g., other utilities, system operators, suppliers, and contractors). Any connected peer must be able to reach any other peer across the network.

On the operational side, a wide range of legacy communication services are required for connecting devices and operators in electrical grid sites (e.g., substations) or to central platforms (control, monitoring, asset management . . .). A large number of new or already deployed applications in the electrical grid increasingly require IP or Ethernet connectivity.

Moreover, applications are substantially growing at the interface between the enterprise and the operational worlds requiring information exchange between field operational sites and utility offices or enterprise information platforms. Some typical cases are described in part 1 of the present book.

It is therefore understandable that decision-makers examine the opportunity of unifying the two networks for cost optimization and for information interchange considering the present similarity in the employed technologies and required skills.

On the other hand, network structures, performance metrics, security constraints, management processes and operation practices of the IT world do not always match the realities and requirements of the utilities' operational applications with over-cost and/or under-performance being typical consequences. The main issues of focus are fundamentally different in the two networks rendering a converged network a very challenging concept to implement.

© Springer International Publishing Switzerland 2017

C. Samitier (ed.), *Utility Communication Networks and Services*,
CIGRE Green Books, DOI 10.1007/978-3-319-40283-3_20

In this section, we intend to explore some specific aspects, to identify some major risks, and to propose some service delivery scenarios for aggregating communication traffic flows in the power utility.

IT and OT diverging requirements

For many years now, enterprise networking in the power utility (as in any other corporate office environment) has entirely evolved into IP service integration. The communications of the electrical grid is a totally different story: many different new and legacy applications with specific interfacing, QoS imperatives, time-sensitivities, and security constraints lead to multiple IP and non-IP operational traffic flows which need to be aggregated together at the Service Delivery Access Point and transported across the grid.

Some notable differences are listed below and summarized in Fig. 20.1:

- Legacy services—These are abundant in the operational network (in particular when dealing with power transmission grids) and a determining factor in the selection of technology and architecture. They are practically inexistent in the enterprise network where all required services are basically IP and favors "feature-rich" networking.
- Structure and Traffic—The operational network is characterized with a large number of sites (electrical substations) with a moderate volume of operational traffic in even in the most ambitious scenarios. The enterprise network, on the other hand, is composed of a small number of sites (utility offices) with

Requirement	Utility Operational Communications	Enterprise Data Communications
Legacy services	Extensive legacy services and devices Must integrate new IP services	Services are (almost) exclusively IP (no legacy integration)
Structure and Traffic	Large number of sites (HV substations) with small volume of data exchange	Small number of sites (utility offices) but extensive data exchange volume
Quality focus	Delay, Service continuity Fault Tolerance	Throughput , Flexibility, Cost and Efficiency
Communication Peers	Mostly pre-determined peers	Any-to any
Life-cycle	Need service stability (Evolves with power applications)	Need frequent network updates (Evolves with IT evolution)
Management	Need fast fault recovery and continuous monitoring	Need frequent service provisioning and configuration change
Outsourcing	Mainly utility-owned network infrastructure in-house operation	External service provisioning and outsourced O&M is common
Security focus	Service availability	Service integrity and confidentiality

Fig. 20.1 Enterprise (IT) and operational (OT) networks requirements

substantial volume of ever increasing sporadic traffic. An integrated approach results in the deployment of complex and costly equipment in substations greatly over-dimensioned for the needs of the substations but often not adapted for the environment.

- Quality focus—Delivering low delay and high service continuity (availability, fault-tolerance, and power-autonomy) are the main focus of the operational network, while an enterprise network gives priority to throughput, flexibility, bandwidth efficiency and cost.
- Communication peers—Mostly fixed and preestablished in the operational network: even field sites access from the office environment (e.g., maintenance centers) is established through a unique and predetermined service access point on the operational network to allow security filtering and authentication.
- Life-cycle—Unlike the enterprise network which is continuously evolving, the creation and modification of connections is relatively rare in the operational network. For many operational applications network change implies application re-commissioning, while an enterprise network, on the other hand, cannot accept a "frozen" network.
- Management process—The operational network requires fast fault recovery and continuous monitoring for essentially predetermined communication peers, while the enterprise network gives priority to frequent service provisioning and change for mainly any-to-any communications. The relatively simple and "fast reaction" O&M process for operating a network of substations is fundamentally different from the more formal management of an enterprise network generally associated with the IT system.
- Outsourcing—In many cases the enterprise communication service is provisioned externally (e.g., through a telecom service provider) or operated and maintained through an external entity. In the power system operational network context, however, the largely dominant mode of operation remains a fully in-house organization and process due to prompt service imperatives (a contractor SLA with "commercial sanctions" not assuring prompt action) and substation access constraints (e.g., requiring identified, "safety trained" and electrically certified intervention staff.
- Security Management—Security assurance mechanisms in enterprise and operational networks are significantly different: pre-determined or any-to-any traffic, closed or open to public network, time sensitive device-to-device or mainly transactional, giving priority to service availability or to confidentiality and integrity.

Mutualizing resources between the two networks

The above-described diverging attributes and requirements of the operational and enterprise networks show the difficulty of unifying the two networks without being a handicap in performance, stability, management process, evolution, or cost. What could then be mutualized between the two networks?

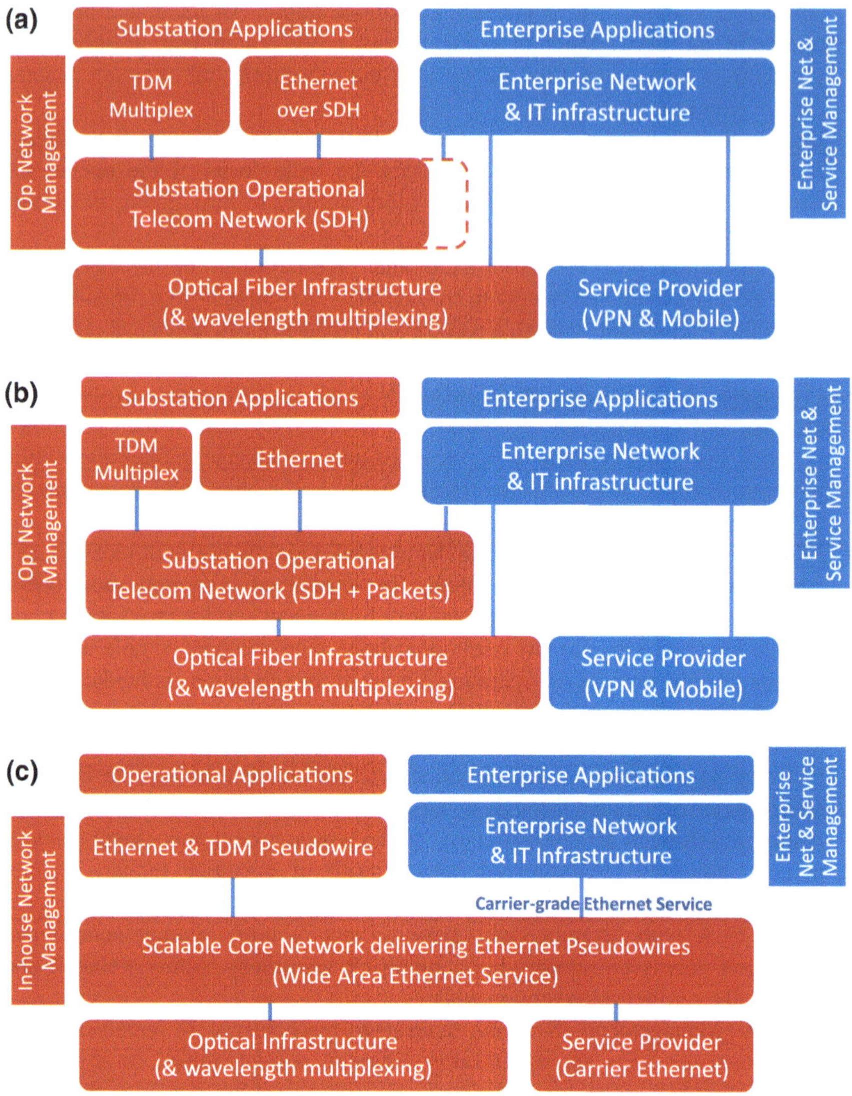

Fig. 20.2 Possible evolution of operational (OT) and enterprise (IT) communication service delivery. **a** Present network overlay and IT/OT network and service situation in a large number of utilities. **b** Short-term evolution with increase of packet-switched operational communications. **c** Possible long-term evolution leading to a common communication resource provisioning platform

A hierarchical architecture comprising a static core transport network and multiple access/distribution layers allows a convergent core transport service of Operational and Enterprise networks without compromising the performance or the required management process of either network: the core provides, in this way, high capacity point-to-point trunk services to the Enterprise network without interfering into its architecture, employed technologies and management.

Figure 20.2 presents the substation operational and corporate enterprise networks distinguishing the current situation (a) in most utilities as well as their possible evolution (b and c) where the core network built for operational usage can deliver connectivity to the enterprise network without interfering with its management and governance.

The transport service delivered in this manner can fulfill all operational requirements and moreover provide carrier-grade Ethernet transport services to distinct corporate enterprise routers without impacting their architecture, technology choices, life-cycles issues and network management. Dynamic routing and allocation can overlay this static infrastructure whenever necessary and managed independently in the operational and in the enterprise networks.

Managing TDM and packet-switched technologies in this integrated manner is the natural continuity of managing Ethernet over SDH (EOS) largely deployed in numerous power utility operational networks today and therefore does not require radical changes in the way the network is being managed, necessary skills, processes, and organization.

To conclude on this issue, one can say that the optimal level of resource sharing in power utilities communication networks is something to be assessed in each particular case: from fully distinct network infrastructures and provisioning modes or sharing of optical infrastructure (dark fiber or wavelength) to common provision of carrier-grade Ethernet transport, different levels of resource sharing can be implemented without penalizing either usage. It must, however, be kept in mind that integration is not an objective on its own: it may mutualize costs of deployment, management and maintenance, and it may facilitate interactions between users. But at the same time, it may increase risks of service outage and security vulnerability, constrain the process and organization of network providers and users, and necessitate common planning, task synchronization and change management as well as extra effort to assure the Quality of Service.

Part IV
Deploying Reliable and Secure Network Infrastructures

21 Deploying Reliable and Secure Network Infrastructures

Delivering communication services to operational applications, as described in part 3 requires implementing a suitable telecommunication infrastructure with adequate capacity but also with a high degree of reliability and security. Such a network must cover all service access points in electrical power substations, control and monitoring platforms and must assure access to and from the technical offices of the power utility. The present part's focus is on planning, designing, and implementing the network infrastructure, as well as the tools, processes, and skills necessary for delivering the required communication services.

This is a wide scope, very far from that of conventional telecommunication textbooks and touches many domains: telecom technology indeed, but also architectural design, different asset ownership situations encountered in the power utility context and their impact on network design, as well as cyber-security and disaster resistance constraints.

Migration is a key word in this context

- User applications and their respective service requirements in the utility evolving slowly, the concept of legacy interfacing and "co-existence" becomes a fundamental one.
- Network technology is in permanent change. Network structures and architectural choices must allow the evolution of telecom technologies and their gradual introduction across the network.
- Skills and capabilities of the utility staff cannot be adjusted spontaneously. They must be planned at the same time as network evolutions and build up gradually with hands-on experience, enhancing utility's capabilities. Tools and processes also can only change in an evolutionary manner, migrating into more adapted schemes if major operational anomalies are to be avoided.

© Springer International Publishing Switzerland 2017

C. Samitier (ed.), *Utility Communication Networks and Services*,
CIGRE Green Books, DOI 10.1007/978-3-319-40283-3_21

The present part begins with a short overview on network technologies to describe some concepts and elements of vocabulary already used in the book (not so easy to put everything in a logical sequence). The architecture of the telecom network is then discussed, introducing the benefits of a hierarchical, overlay structure. Upstream management, introduced in the previous part is discussed in more detail, providing some insight into long and medium term network planning, the deployment of capabilities, and short-term network adjustments. The concept of asset ownership is the subject of a specific section to describe its impacts in the new deregulated utility context. The discussion follows into the subject of migrating from circuit to packet with its different approaches. Finally, some specific network implementation issues comprising cyber-security and service continuity are discussed.

22 An Overview on Network Technologies

Power utilities employ a wide variety of telecom technologies in their operational communication networks. These technologies, which evolve over time, are extensively documented in general telecom literature and their detailed analysis is beyond our present scope. A short overview of some major ones used at present is given here. Copper wires are not considered in this work and PLC is only analyzed as far as new technical advances are concerned. The concepts and terminologies introduced in this section are used across the book, even in preceding parts.

It is also necessary to introduce the different modes of sharing common communication resources that are employed in power utilities to integrate operational (and sometimes non-operational) services. This includes multiplexing of dedicated frequency bands, optical wavelengths, time intervals in a high-speed data stream, or dedicated packets in a high capacity data link.

22.1 Multiplexing and Switching Fundamentals

Wide Area Telecommunication Networks carrying operational services for electric utilities are commonly based at present on Circuit Switched technology, where a connection between devices is established ("switched") using a fixed path (a "circuit") through the network. This path—once established—offers a fixed bandwidth and a fixed and deterministic latency to the service. For the simultaneous transmission of a multitude of services (i.e., many data, voice or video signals) sharing the same medium, these signals are combined using **Time Division Multiplexing (TDM)**, where each signal gets cyclically allocated a time slot with a time gap short enough to be unnoticed by the user.

The disadvantage of this technology is that the network resources are inefficiently used by sporadic, burst traffic like voice or video. For example, the bandwidth allocated to a voice channel is permanently occupied even during speech pauses or when not used at all. The advantage is that each service receives its dedicated network resource ("channel") with fixed bandwidth and stable end-to-end

© Springer International Publishing Switzerland 2017

C. Samitier (ed.), *Utility Communication Networks and Services*,
CIGRE Green Books, DOI 10.1007/978-3-319-40283-3_22

delay. Congestion by competing traffic and time-irregular delivery of the service are avoided. PDH and SDH technologies described in the following sections are examples of TDM networks.

In opposition to TDM, **Frequency Division Multiplexing** (**FDM**) provides a separate frequency band allocated to each service as used in Analogue Power Line Carriers and in radio systems. The same principles applies to **Wavelength Division Multiplexing** (**WDM**) employed over optical fibers. In these systems, the interaction between different services is only through potential cross-band interferences. It should be noted that resource allocation through TDM and FDM (or WDM) can be simultaneously used in different hybrid configurations.

The disadvantage of poorly exploiting network resources has been tackled by the so called **Packet-Switched** technology. Simply speaking, the data to be transmitted is split into packets (which can be of various sizes) and are labeled with a source and destination address. The packets enter a queue at the ingress of the network, and as network resources become available the packets are forwarded by switches and routers through the network, normally following the shortest or the fastest available path. This technology is employed in Wide Area Networks (WAN) based on Ethernet or IP packets.

Packetization, queuing, and competing for network resources (bandwidth), inherently generate the risk of a nondeterministic service delivery which is a potential problem for time-critical signals such as protection. Bandwidth over-provisioning combined with advanced techniques such as service prioritization, traffic engineering and bandwidth reservation, time distribution protocols, data buffering and Synchronous Ethernet contribute to gradually overcome or reduce most of the undesired statistical behavior of packet-switched technologies.

22.2 Optical Communication

Optical fibers are widely deployed in power utility communication networks and constitute the preferred communication medium between substations as well as for the substations' internal automation networking. Fibers can be dedicated to a single application (e.g., protection relays), or may be shared for the usage of many applications.

Power utility communication networks employ exclusively Single Mode (SM) fibers, allowing better transmission characteristics such are attenuation, dispersion and bandwidth. Multimode (MM) fibers are limited in their application to short distances inside the substation (protection relays, actuators, metering devices, etc.) to assure galvanic isolation at low cost of installation and maintenance.

Power utility optical fiber cables are generally deployed along the overhead, underground, or undersea power lines. The fibers are illuminated by laser emitters and can be used in single wavelength or multiple wavelength modes. In the latter case, described in the following section, every wavelength acts as an independent carrier of information increasing the fiber capacity or reinforcing service separation.

Many optical cable technologies are presently in use depending on the nature and voltage level of the power transmission lines. In the case of overhead power transmission lines, the optical cable may be incorporated into the protection ground wire OPGW, or wrapped around it.

Alternatively, the optical cable may be "self-supported," attached to transmission towers or distribution poles. All Dielectric Self-Supporting (ADSS) optical cables are generally used in existing line retrofit when the line voltage is below 150 kV, and where the line is not equipped with protective ground wires with full continuity. This is often the case of medium voltage distribution lines.

However, above 150 kV, the presence of electric field *E* affects the transmission characteristics of the fiber increasing its birefringence effects. Moreover, at higher voltages, ADSS is generally not employed due to the effects of the electrical field on the dielectric sheath. For these systems, the preferred technology is OPGW with the optical fibers placed inside steel, aluminum or plastic tubes to protect 12, 24, 48, and 96 optical fibers according to different cable designs.

The available capacity of optical fibers is often assumed to be almost limitless based on advances in optical communications although factors such as temperature change and strong electrical fields may limit these capabilities at very high communication speeds (e.g., 40–100 Gbps Ethernet).

22.3 Wavelength Division Multiplexing (C- and D-WDM)

Wavelength Division Multiplexing (WDM) is increasingly in use in power utility communication networks to provide separation between different types of services (operational and corporate networks, legacy, and new packet core network, fiber-based protection relay and other communications, etc.). This is particularly true where extra fiber is not available or cannot be provisioned economically.

WDM allows multiple optical beams from different optical communication systems, each at a different optical wavelength, to be combined through an optical multiplexer and injected into the same fiber. At the reception side, the different wavelengths are separated by a passive optical filter bank and treated by different optical receivers.

WDM provides full isolation of telecom services which only share the fiber and possible optical amplification.

Coarse Wavelength Division Multiplexing (CWDM) is defined by ITU-T Rec. G.671 standard. It allows implementing reliable and cost-effective systems, through a combination of uncooled lasers, relaxed laser wavelength selection tolerances and wide pass-band filters. CWDM systems can be used in transport networks for a variety of architectures, delivering a very cost-effective solution. Figure 22.1 presents the channeling structure for CWDM together with fiber attenuations.

It is worth noting that the water absorption peak around 1400 nm is no longer present in more recent fiber designs (ITU-T G.652-C or G.652-D) in which the OH ions inside the crystal structure have been removed through a modified manufacturing process.

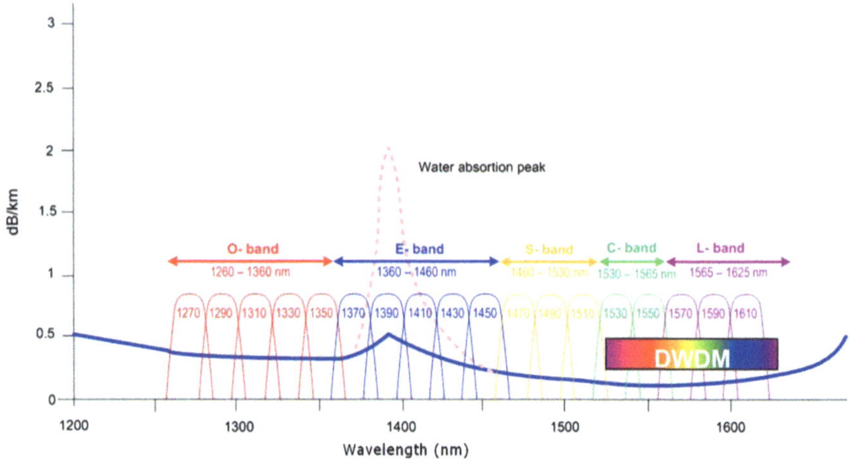

Fig. 22.1 Channel structure for CWDM

Dense Wavelength Division Multiplexer (DWDM) on the other hand, is a technology used for multi-channel long distance and/or very high capacity links (ITU-T G.694.1). DWDM systems, mainly used by Public Telecom Operators, can multiplex from 32 to more than 100 channels in the range of 1530–1624 nm. With a channel spacing of 1 nm (or below), DWDM operation requires extremely precise optical sources (laser temperature stabilization) resulting in significantly higher complexity and cost.

With a channel spacing which is 10 times larger, CWDM systems tolerate the temperature drift over the whole industrial temperature range of the laser emitters without leaving their allocated channel. This ruggedness, in addition to considerably lower cost, constitute significant advantages in the Utility communication network where wavelength multiplexing is not often used for attaining maximum bandwidth but rather for separating networks.

For using optical amplification systems in long spans, CWDM systems should be limited to four channels (C- and L-bands) to make them compatible to the limited bandwidth of these devices.

22.4 Time Division Multiplexing (PDH and SDH)

The great majority of power utility telecom networks are presently composed of digital **Time Division Multiplex** (**TDM**) systems. According to the system's generation, capacity and followed national/international standards, these are called Primary Multiplex System (E1/T1), PDH, SDH, SONET Channel Bank, Add-Drop Multiplexer, Digital Cross-connect, etc.

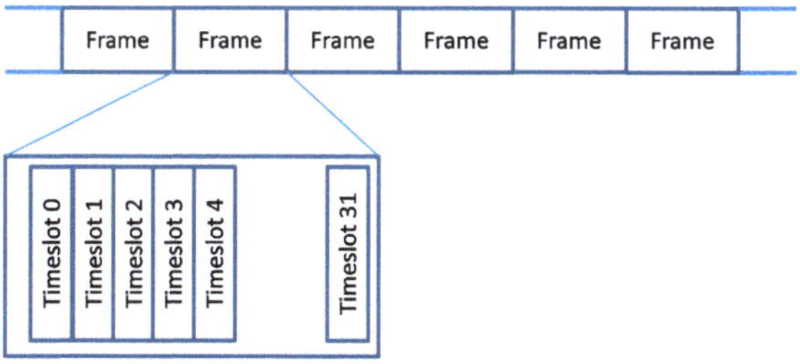

Fig. 22.2 E1 frame: 32 timeslots per frame, 8000 frames per second

Operational networks extensively use these systems at present. It is therefore necessary to give a more detailed introductory coverage of the operating principles involved.

Historically, in order to exploit the digital transmission capacity of wired or wireless transmission systems, the digital bit-stream has been used to transport continuously cyclic frames of fixed length which are partitioned into equal **Time Intervals** (**TI**). Using the same TI in each frame, one can transport an elementary digital stream, time domain multiplexed with other digital streams. A frame header sequence allows synchronizing both ends of the link and therefore to separate multiplexed characters at the receiving end.

Historically, the **elementary digital stream** has been fixed at 64 kbps (56 kbps in North America) corresponding to 8000 analogue samples/s with each sample coded over 8 bits (8000 sample/s required to convert a 4 kHz analogue voice signal to digital and reconvert back to analogue after transmission across a link without loss of information).

Time multiplexing of 32 elementary digital streams (30 useful streams plus slot 1 for Header sequence and slot 16 normally reserved for signaling) has yielded a multiplex system with an aggregate bitrate of 2048 kbps (currently referred to as 2 Mbps or **E1**) shown in Fig. 22.2. Slightly different constitution in the North American and Japanese systems gives a 24 channel multiplex at 1544 kbps (referred to as **T1 and J1**).

The E1 system is byte-multiplexed (8 bits per time interval), has a frame length of 256 bits, and a periodicity of 125 μs. This system is referred to as **Primary Multiplex System** (ITU-T G.703) and provides analogue VF interfaces as well as low-speed synchronous and asynchronous data interfaces ranging from 300 bps to $N \times 64$ kbps. A primary multiplex system is purely synchronous: at the receiving end, the system must reconstitute clocks from the received signal for bit reception and for frame synchronization. Frame synchronization is performed through initial searching of the frame header sequence and multiple successive reception of the same pattern at the right time. Synchronization is lost when the pattern is not found at the right place a determined number of times and a new search for header is initialized. The loss of synchronization and the consequent time of resynchronization can lead to periods of unavailability which can impact the performance of certain utility operational applications such as protection signaling if no special measure is taken.

It should be noted that the primary multiplexer performs the sampling of analogue signals and low speed asynchronous serial data and therefore yields purely synchronous elementary digital streams. Similarly, when external data streams at 64 kbps or less are to be transported, adequate clock exchange mechanisms exist at the data interface to assure the synchronization of all elementary streams of the multiplexed link.

However, when multiple E1 streams from different primary multiplexers are to be combined together to use a higher capacity transmission resource, the clocks of the different primary multiplex systems, although at the same nominal frequency, are never identical. Clocks at the same nominal frequency are called **Plesiochronous**.

Higher level combination of primary multiplex data streams is performed through bit multiplexing, (combine four 2 Mbps to constitute an 8 Mbps stream and then combine by four to constitute a 34 Mbps, etc.). Successive multiplexing produces a **Digital Hierarchy**.

In order to overcome the slight difference in clock speeds of the constituent bitstreams, the fastest clock is adopted and extra bits are added to the slower streams at predetermined positions (bit-stuffing) to avoid "buffer underflow." These bits are identified and extracted at the de-multiplexer. This digital hierarchy of non-identical rate (but nominally equal) bit streams is known as **Plesiochronous Digital Hierarchy** (**PDH**).

Primary Multiplexing is the basic signal aggregation mechanism in the great majority of Utility networks (now complemented by Ethernet switches). PDH links at 8 and 34 Mbps are commonly used in power utilities over optical fibers and over microwave radio.

PDH multiplexing has been devised (and was well-suited) for the connection of Telecom Switching Centers (Central Office, CO). It is not appropriate for Adding and Dropping of "containers" at intermediate stations across the network. This means that the whole load of channel groups in the hierarchy needed to be "unfolded" whenever some load needed to be "dropped or added". This method results in unreasonable time delay, complexity, costly equipment and excessive cross-connect wiring.

Synchronous Digital Hierarchy (**SDH**) and its US counterpart **SONET** (**Synchronous Optical Network**) replace this cascaded folding/unfolding mechanism. The following description refers more specifically to SDH although the principles apply equally to SONET systems used in North America.

The Synchronous Digital Hierarchy is composed of multiplexed digital rates of 155 Mbps (STM-1), 622 Mbps (STM-4), 2.5 Gbps (STM-16) and 10 Gbps (STM-64). Power utility telecom networks employ currently STM-1, STM-4, and STM-16.

Unlike PDH, SDH is a network-based system using **Add-Drop Multiplexers** (**ADM**) at each node loading and unloading data streams in previously reserved **Virtual Containers** (**VC**) for connecting to different destinations. The available data capacity is partitioned into a Pointer domain and a Payload domain. Each VC in the SDH payload corresponds to a specific pointer indicating the location of the VC in the payload. The VCs in the same payload can have different sizes (number of reserved bytes in the payload) corresponding to the capacity of the data stream (2, 34 Mbps, etc.). The minimal size of the VC is 2 Mbps (VC12) allowing the transport of an E1.

Different levels of overhead (segment, path, etc.) allow more elaborate management and end-to-end recovery of connections across the network. In particular, **SNCP** (**subnetwork connection protection**) protects in a differentiated manner each tributary data stream through an alternate path around a ring.

Transport of data across an SDH network is performed through previously reserved capacity and centrally controlled deterministic routing leading to a highly predictable transmission delay. Furthermore, the management information accompanying each Virtual Container enables very fast data traffic recovery on network fault conditions (<50 ms).

However, the self-healing mechanisms of SDH are not implicitly bi-directional and may lead to different go and return paths and consequently delay differences causing anomalous operation of Differential Line Relays or other device using an echo method for time synchronization. Precise network design or specific mechanisms are required to avoid such asymmetry in the network.

SDH systems deployed in the 1990s were focused on the transport of circuit switched traffic (E1, etc.) with only limited and proprietary facilities for Ethernet packet transport. New Generation SDH (NG-SDH) has since been completed with a set of standards for optimized mapping and control of Ethernet traffic (**Ethernet over SDH or EoS**) as described in the part 3. This allows power utilities to combine at an improved bandwidth granularity, the more efficient and flexible bandwidth sharing of packet networks with the deterministic behavior, low delay, and fast healing of SDH.

22.5 Optical Transport Networks (OTN)

Optical Transport Network (OTN) is an emerging flexible technology based on SDH/SONET adapting to the great increase of data traffic in transport infrastructures. It is described in general terms in ITU-T G.872 and the network interface is specified in ITU-T G.709. OTN is a highly scalable core network technology capable of time multiplexing the existing SDH stream with packet-switched data over the same frame. SDH/SONET, Ethernet, Synchronous Ethernet (SyncE), IP-MPLS, and MPLS-TP can all be accommodated by OTN. This technology provides a deterministic behavior similar to SDH/SONET.

The different constituents of the OTN hierarchy are presented in the table hereafter.

OTN hierarchy	Capacity (Gbps)	Transport capability
ODU-0	1.25	STM-1, STM-4, GbE
ODU-1	2.5	STM-16, GbE
ODU-2	10	STM-64, 10GE
ODU-3	40	STM-256, 40GE
ODU-4	100	STM-64, 100GE

The communication capacity of OTN is well beyond the estimated requirements of utility's operational applications but is used in power utilities for simultaneous transport of both operational and enterprise networks at the core level. The higher elements of the hierarchy, ODU-3 and ODU-4 use Dense Wavelength Division Multiplexing (DWDM) for providing capacities up to 100 Gbps.

Optical transport network constitutes a higher layer of transport network to supplement existing SDH networks for higher capacity and in particular for integrating large volumes of packet-switched traffic. It does not, however, support lower speed interfaces such as E1 or sub-E1 legacy and cannot therefore replace lower level SDH/SONET multiplexing.

The OTN data frame integrates Forward Error Correction (FEC) which allows operating with a higher optical dynamic range and therefore longer spans, in particular when wavelength multiplexing is employed. The G.709 FEC is based on a Reed-Solomon RS (255, 239) code adding 16 bytes of error control to 238 + 1 bytes of payload.

Moreover, the OTN frame includes transport overhead that provides for enhanced operation, administration and maintenance (OAM) capabilities. SDH-like OAM channels provide monitoring facilities which are extended to multiple tandem connections.

22.6 Ethernet Transport

Ethernet is a data link layer (Layer 2 or L2) communication protocol and can be carried over any physical transmission support (Layer 1 or L1) such as an SDH network (EoS), or natively over optical links or microwave radio.

As mentioned in the previous paragraph, SDH has been complemented with specific protocols to integrate Ethernet packet traffic to the VC constituting the SDH payload. Ethernet encapsulation over SDH is defined by Generic Framing Protocol (GFP), Virtual Concatenation (VCAT) and Link Capacity Adjustment Scheme (LCAS) already described in part 3. However, matching, asynchronous data streams such as Ethernet packets to a constant rate container system such as SDH becomes complicated, inefficient, and expensive as the proportion of packet to TDM traffic grows.

Native optical Ethernet transport between substations becomes a candidate solution as the extent of legacy traffic decreases in quantity and importance as compared to natively Ethernet-interfaced applications. This is indeed to be thought of, in the present context of IEC61850 substation communication architecture and gradual migration of substation Intelligent Electronic Device (IED) into a networked environment. Native optical Ethernet can use dedicated fibers or dedicated wavelengths (C- or D-WDM).
Native optical Ethernet transport, however, suffers a number of shortcomings

- Ethernet protocols include mechanisms for restoring connectivity across a meshed network of switches when one or multiple nodes or links become unavailable due to faults (Spanning Tree Protocol). The time needed for such restoration, however is in the range of seconds, considerably longer than SDH (<50 ms).
- Aggregating of Services and Delay issues—When multiple communication services are to be transported together, TDM (e.g., SDH) allocates predetermined communication capacity to each application, guaranteeing therefore a fixed and predetermined transfer time across the network (L1 separation). Native Ethernet, on the other hand, uses the principle of packet queuing. Data packets belonging to different applications are stored in a queue and processed sequentially at each node. Ethernet switches may have the capability to distinguish packets belonging to different applications by associating specific tags to data frames belonging to each connection (L2 separation). Each group of similarly tagged frames constitutes a **Virtual Local Area Network** (**VLAN**). Different levels of priority (different queues) may be assigned to different VLANs (IEEE802.1 P and Q). However, priority assignment does not guarantee fixed delay.
- Scalability—As the number of virtual connections to be deployed across the network grows, VLAN allocation proves to be complex and not scalable. Cascaded tagging (stacked tagging) mechanisms have been added to Ethernet to overcome this issue. IEEE 802.1ah, also called Mac-in-Mac, is a native Ethernet mechanism (L2) which overcomes the scaling limitation of VLANs by re-encapsulating the traffic with an outer Ethernet header carrying input and output addresses of the Ethernet transport network. In this way, the provider's Ethernet switches no longer need to learn the large number of end-user MAC addresses. This technique called Provider Backbone Bridging separates Ethernet transport from Ethernet LANs of the end-users.

22.7 Multi-protocol Label Switching (MPLS)

Multi-Protocol Label Switching (MPLS) is a packet forwarding technique which was specified by the Internet Engineering Task Force (IETF) in the 1990s in order to introduce ATM-type "connection-oriented" traffic handling and Quality of

Service independent of IP destination addresses in connectionless, essentially best effort IP networks operating with Datagrams and using IP addresses for forwarding packets.

MPLS operates through the insertion of a label at the Ingress port of the network and associating a Forwarding Equivalence Class (FEC) to the label. Each FEC defines the manner in which the packet is to be treated in each MPLS node. The label is removed at the Egress port of the network. A connection established in this way is called a Label Switched Path (LSP).

Labels are swapped at intermediate nodes according to a plan defined through a distributed control plane using a Label Distribution Protocol (LDP). This exchange of control information between the network nodes each of which taking part in the definition of paths is a great force giving scalability to very large systems but also a source of substantial complexity which may often be found unnecessary in the scale and size of an electrical power utility dedicated network.

IP-MPLS

The MPLS labels are replaced at each switching node. This provides high flexibility and dynamic routing capability but it requires network-wide intelligence, implying permanent exchange of control information across the network. This is known as a "distributed control plane." Moreover, in order to enable each node to allocate capacity for each traffic stream some traffic rule or policy must be given to the node. A distributed control plane generates critical control traffic across the network and traffic engineering implies some prior knowledge of traffic classes and attributes. The IP-protocol control traffic is transported using a resource reservation protocol RSVP. Such an operation is suitable for extremely large networks with large variations of traffic volumes and source-to-destination paths but well characterized traffic models. This is typically the case for Telecom operator networks. Adjusting Traffic Engineering parameters in such a system is complex but far less laborious than configuring thousands of predetermined data paths which moreover can change with time of day and with seasons according to network users' bulk movements. The technique is known as MPLS-TE (TE means Traffic Engineering).

IP-MPLS, in its simplest form, uses no traffic engineering but simple Priority Assignment as in native Ethernet. The most time-sensitive traffic will only have a better delay variation than lower priority traffic. Such a system provides traffic isolation advantages of MPLS, dynamic routing through a distributed control plane but no guarantee on resource allocation to each service (except for the control plane exchanges using RSVP). It suits multi-site enterprise networks such as utility non-operational networks with no tight time and Quality of Service imperatives.

MPLS-TP

Transport Profile MPLS (or MPLS-TP) is a variant of the MPLS technology simplified for transport-type networks providing predefined tunnels between Service Access Points. It is an adaptation of MPLS to support traditional transport network requirements such as high availability and QoS support, presently fulfilled by SDH/SONET transport technologies.

MPLS-TP data traffic is forwarded across the network according to labels which are added at the network ingress and removed at the network egress point without being changed at intermediate switching nodes. This provides a predetermined tunnel with no "distributed control plane."

Packet forwarding tables at each node are provided by a Central Control Platform in the same way as in SDH. The centralized control, moreover allows predetermined end-to-end alternate routes, and simple resource allocation. The control platform can allocate a maximal capacity (Peak Information Rate or PIR) and if necessary a minimal capacity (Committed Information Rate or CIR) to each data stream, hence assuring SDH-like Quality of Service for time-critical services and IP-type flexible bandwidth for non-time-critical services.

MPLS-TP decouples the "end user-to-end user" network into distinct segments of Customer Network (Application Network) and Transport Network. The latter segment requires essentially point-to-point services known as "Pseudo-wires" (PW) transporting Ethernet (E-line or E-LAN Carrier Ethernet services) and legacy TDM circuits upon the packet-switched network (TDM Pseudowire Emulation).

Moreover, the focus on transport domain allows MPLS-TP to use only a subset of MPLS, removing nonrelevant features and their corresponding complexity while adding some missing functions necessary for transport networks and existing in SDH/SONET such as implicit OAM signals exchanged at different levels of connectivity.

It is clear that MPLS-TP cannot suit a large telecom operator network covering final customers due to its fixed "tunnel" approach (transport profile) and can become laborious for very large numbers of end points and services encountered in the public networks. For electrical power utilities, however, MPLS-TP seems a very promising packet-switched network technology, providing all features of the existing SDH/SONET system, compatible with enterprise level core networks (e.g., IP-MPLS), and delivering new capabilities with similar management processes and skills as the ones already existing in the power company (Figs. 22.3 and 22.4).

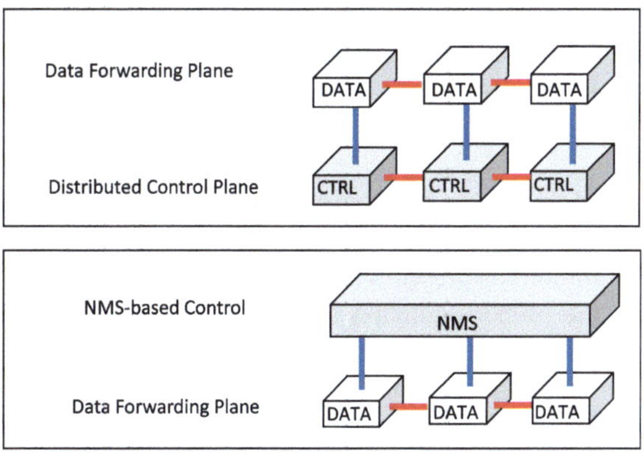

Fig. 22.3 Data and control planes in IP-MPLS and MPLS-TP

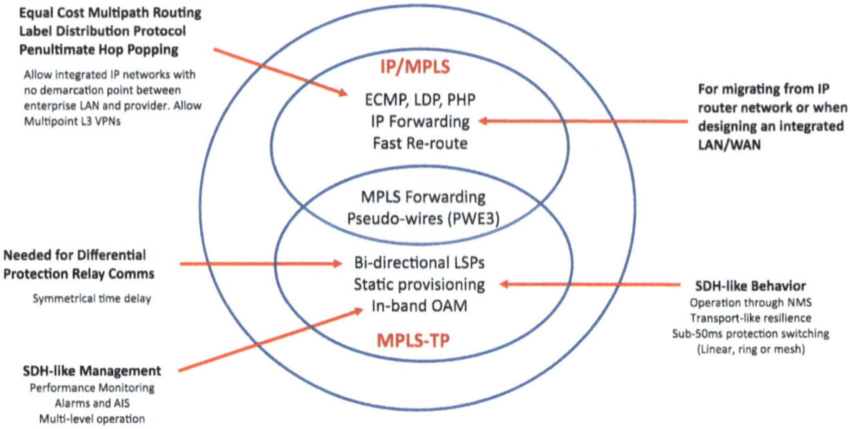

Fig. 22.4 Functional differences between IP-MPLS and MPLS-TP

22.8 MPLS-TP or IP-MPLS in Operational Context

Deploying IP-MPLS or MPLS-TP in the power utility operational network is at present a subject of debate in the industry. Equipment designs seek to obtain the better of two worlds through specific management tools to simplify the implementation of IP-MPLS, or through integrating L3 interface capabilities into MPLS-TP platforms. The lack of maturity of MPLS-TP due to its much later introduction is also being absorbed in time. The present section is therefore only a temporary assessment, made at a given instant of time.

In the power utility context, deploying MPLS is not a "build from scratch" but part of an evolutionary transformation plan as discussed in Sect. 26 hereafter. As already stipulated in the introductory section, "migration" is a key word in deciding upon technologies. The present network is often SDH or SONET, time-sensitive legacy services are designed for circuit mode operation, and synchronous data streams such as E1 often need to be emulated across the packet core. Moreover, the utility controls its telecom network from a central network management platform with processes and skills which are adapted to such operation. MPLS-TP in this context seems to be a more appropriate technology than IP-MPLS with or without traffic engineering. Some of the main reasons are summarized below

- **Maintaining full Control of the Network**—In MPLS-TP forwarding labels are produced by a central NMS allowing end-to-end main and alternate route definitions as presently done through SDH/SONET. In IP-MPLS on the other hand, the proper operation of the network depends upon control plane communications. Moreover, if a deterministic behavior is necessary for some data streams, then adequate traffic engineering must be introduced to govern nodes decision. Adjusting traffic engineering parameters in an IP-MPLS network (MPLS-TE) is complex and requires tuning and adjustments rendering the network subject to non-optimal settings.

- **Quality of Service and deterministic behavior**—Some utilities using dedicated resources such as separate fibers for Protection may imagine that these constraints are definitively out of their shared network. In reality, other protection-like applications with network-wide coverage will be deployed in the coming years and dedicated fiber used between adjacent substations can no longer be the solution (e.g., System Integrity Protection Schemes SIPS). Being able to implement time-sensitive applications remains a "must" in an operational network.
- **Size of network and type of traffic**—By its SDH-like behavior, MPLS-TP responds to all existing service requirements, as well as new packet-based services in utility-sized networks. Deploying MPLS-TE on the other hand, is suitable for public networks with highly dynamic data traffic characteristics and too many nodes for centralized control: traffic rules are hence given to nodes so that they can build the forwarding labels at any moment (TE).
- **Capability versus complexity**—IP-MPLS provides numerous technical capabilities but with increasing complexity. Implementing QoS through Dynamic Resource Reservation (RSVP) can be done in IP-MPLS but making it for hundreds of connections is far from being trivial. This level of dynamic complexity for a "static by substance" service seems unnecessary. Similarly, performing any meaningful Traffic Engineering in IP-MPLS (MPLS-TE) requires fairly good knowledge of the traffic shapes and characteristics which is far from being the case for a lot of new coming and future services over the operational network.
- **Migration from SDH**—Transition to packet for power utility networks having extensive SDH infrastructure, management tools and skills is almost smooth for MPLS-TP because of its SDH-like behavior and network management. IP-MPLS is a jump into another type of network operation. With a large telecom network, utilities will become entirely dependent on their supplier. IP-MPLS suppliers provide specific tools and features to overcome some of the basic problems but in a nonstandard manner causing further dependence on a single supplier.

22.9 Radio Communication

Radio communication covers a very large spectrum of systems and solutions used in power utilities to various extents. In general, they present the advantages of fast deployment where no existing wires can be used and the potential mobility of the communicating peers. On the other hand, they have implicit limitations in service availability (propagation conditions, obstacles, fading due to multipath propagation, etc.), bandwidth limitations compared to wired solutions, frequency allocation and licensing issues, as well as their associated tower and mast infrastructures to assure long coverage.

Some particularly useful applications of radio communication in the power utility are as follows:

- Line-of-Sight microwave radio systems (2–13 GHz) are often used to complete optical networks for covering spans which are impossible or difficult to implement in optical fibers (e.g., crossing main motorways, rivers, etc.) or connecting the network to non-electrical sites (e.g., Control Centers, Office Buildings, etc.). Frequency bands are increasingly difficult to obtain in the low end and are constantly pushed to 10–13 GHz or higher, with increased difficulty for assuring path clearance (Fresnel zone clearance) and hence requiring more costly tower infrastructure (when not installed on a building top) whose height is proportional to line-of-sight constraints. Microwave links, through their high frequency and high directivity are allocated large bandwidths enabling them to transport high capacity data flows (e.g., 155 Mbps TDM or 100 Mbps Ethernet). However, unlike fiber systems, their capacity cannot be increased with growing requirements beyond the allocated frequency bandwidth. WiMAX technology, defined by IEEE 802.16x standards, is still in use mainly in North America providing a point-to-multipoint microwave solution with around 70 Mbps capacity. Line-of-Sight radio systems are also used in UHF radio bands with lower capacity (according to different national frequency regulations) but lighter infrastructure (less directional antenna systems, partial clearance of Fresnel zones leading to lower antenna heights). These are frequently used in industrial radio systems deployed in distribution grids.
- Broadband packet-based wireless network IEEE 802.11 (WiFi) is mainly used in the distribution grid for providing a high throughput network to cover the emerging smart distribution applications. IEEE 802.11 in particular enables the implementation of Broadband Wireless Mesh infrastructures with ad hoc network configuration algorithms to cover a distribution network perimeter. A network capacity of around 400 Mbps can be attained through aggregating multiple channels. Powerful encryption algorithms overcome the security issues often associated with the usage of wireless in operational networks.
- Narrowband industrial UHF radio systems provide up to around 20 kbps over 12.5 kHz allocated frequency bands. These systems are widely used in distribution grid automation and SCADA applications. In some countries allocated bandwidths of 25 kHz or more allow higher link rates.
- Private Mobile Radio (PMR) also known as Land Mobile Radio (LMR) are UHF/VHF radio systems allowing the communications of utility mobile work force. One or several radio channels are shared among multiple mobile terminals. Different Trunking protocols such as TETRA or MPT-1327 are used by a central station to control channel access, network signaling and some specific closed network functions. Implementing an adequate coverage through a PMR is costly compared to public alternatives considering the area to cover (number

of base stations) and generally small number of mobile terminals using the network. The major advantage over public cellular being the availability of the network in the event of power outage or other disaster situations.

- Private Cellular radio covers different technologies such as GSM-R (used mainly in transportation) and LTE (4G) are trunked cellular solutions that can be implemented in a private utility operations network to provide very high communication capacities. These technologies can find their place in the future smart distribution network integrating a large number of bandwidth-consuming applications (large-scale real-time metering and demand response) although procuring frequency licenses may be hard to justify for a power utility.

22.10 Power Line Carrier

Narrowband Power Line Carrier (PLC) allows the transmission of communication signals, in particular protection relay communications, over HV power transmission lines. This technique, in use for a long time, still provides a dedicated infrastructure fully incorporated into the substation with a signal path that follows implicitly the substation-to-substation connections over long distances without any repeaters (>500 km) and therefore an interesting solution for transmitting little information (e.g., teleprotection signaling, low speed data, and speech) over long distances. Channels for different applications are aggregated together through frequency multiplexing (FDM) and transposed to a carrier frequency (roughly 25–500 kHz) or time multiplexed to construct a digital data stream which is translated in frequency through digital modulation schemes.

Power line carrier is used when no fiber is available or as a second path to secure the operation of a protection system. The availability of the communication channel being strongly correlated with power transmission line fault conditions, PLC is unsuitable for transmitting current differential protection, coded sampled values (e.g., IEC 61850), or digital stream teleprotection signaling (due to potential synchronization loss on line faults). It is employed for transmitting voice-frequency command type teleprotection, speech signals and data.

Power Line Communications is also used in the power distribution grid for building Advanced Metering Infrastructure (AMI) covering smart meters in customer premises, public lighting control, and different distribution automation and SCADA applications. These systems range from lower bandwidth PRIME, G3 and IEEE 1901.2 solutions with a capacity below 120 kbps, to Broadband Power Line (BPL) solutions allowing to building a packet network over the MV or LV distribution network conductors. Different techniques are used for crossing open breakers, for eliminating band-limited noise and disturbances, and for operating over variable grid topology and characteristics (e.g., impedance). The system can typically provide 10–100 Mbps capacity with a span of 3–4 km between nodes. BPL solutions can integrate multiple applications, each of which being allocated a distinct VLAN and may be used for higher data volume and resolution metering applications (e.g., for Demand Response), backhauling of smart metering, grid automation, SCADA and other distribution grid applications.

Hierarchical and Overlay Architectures 23

As discussed in parts 1 and 2, substation to substation applications require information exchange between fixed peers generally adjacent or a small number of links away in terms of telecom network connectivity. They require little or no network routing resilience but tight constraints on time behavior. This pleads in favor of "more physical" communication architectures providing latency control and implicit security. Service continuity in this case is to be assured through fast and simple preprogrammed switching of end-to-end normal and back-up paths.

Network-wide applications, in particular those concerning human operators at undetermined locations and remote processing platforms on the other hand, require flexibility and high-level resilience as well as more elaborate security (access filtering, encryption, authentication, etc.). This pleads in favor of "more logical" communication architectures managing topologies and data routing automatically.

Adopting a hierarchical architecture not only combines these diverging requirements but simplifies network deployment, service migration, trouble-shooting and testing, network extensions and upgrades and future technology migrations. Referring to Fig. 13.2 on Service Delivery Architecture, the Transport Network (TN) can then be transformed as presented in Fig. 23.1, with:

- A substation aggregation level allowing the interconnection of electrical substations for time-critical applications and providing substation access to the high-level transport network
- A core network level providing access to control platforms and network-wide transport to substation aggregation networks

In this book we refer to the interface between the two networks as Core Network Access Point.

The requirements and constraints in the two levels of transport network are different as summarized in the following table presented in Fig. 23.2.

This approach results in a hierarchical network in which different technologies and different evolution plans can be applied to each part. Moreover, the network

© Springer International Publishing Switzerland 2017

C. Samitier (ed.), *Utility Communication Networks and Services*,
CIGRE Green Books, DOI 10.1007/978-3-319-40283-3_23

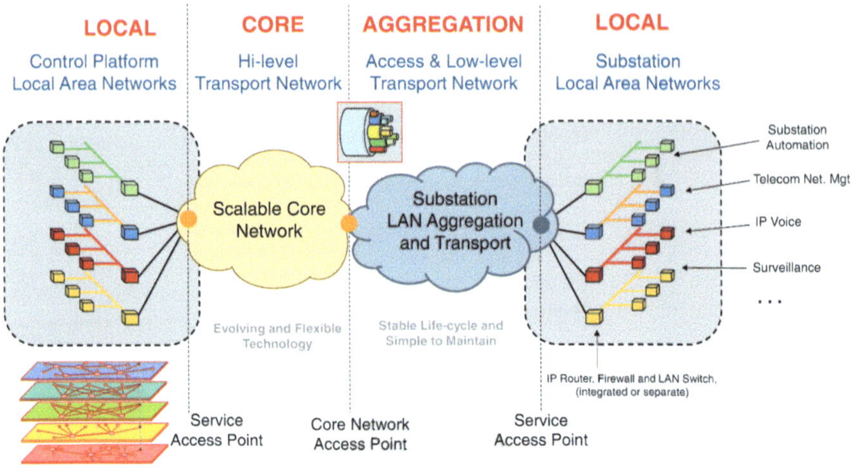

Fig. 23.1 Partitioning of the transport network into substation aggregation and core networks

Substation Aggregation Network	Core Network
Require freezing of configurations to fit long life-cycle power applications such as Protection.	Require more frequent upgrades to keep up with release, bandwidth and technology evolution.
Must be easy to deploy and easy to maintain considering large number of sites and low level of specialized telecom skills of field staff.	Can tolerate more complexity considering the relatively smaller number of sites and usage of more specialized telecom staff.
Migration is extremely slow due to number of sites and applications. Must co-exist with old infrastructure in many sites with gradual service switch-over.	Can be quickly deployed on separate fiber or wavelength and be used as a support structure for the more gradual deployment of substation networks.
Time-critical applications such as Protection are often confined to a same substation network.	Enterprise networks if transported over the same network are confined to core network only.

Fig. 23.2 Substation aggregation and core network requirements

may be further partitioned in larger networks, and different technologies may be used in different substation aggregation networks to build a gradual migration plan or to fulfill diverging requirements (e.g., legacy and networked automation substations) (see Fig. 23.3).

Another concept currently used in power utility telecom network architectures is that of "network overlay". We have already discussed in part 3 and in earlier sections of the present part that legacy circuits, primary multiplexer network access and Ethernet access can all be transported over SDH, over native Ethernet transport, and over MPLS, each of which, just as each of the former user interfaces, can be transported over fiber, radio and (for some) over narrowband media.

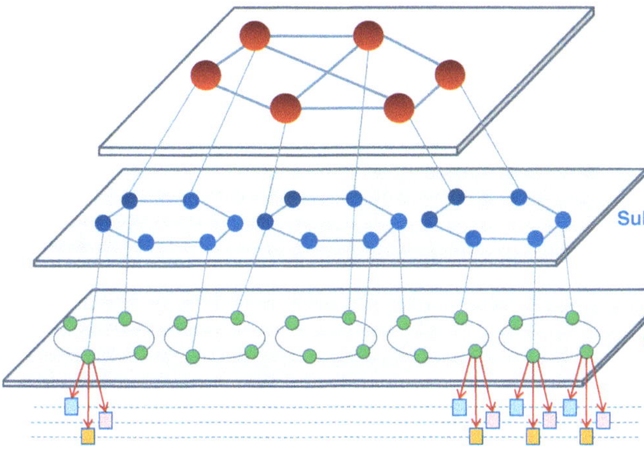

Fig. 23.3 Multi-level hierarchical transport network

In practice, it is rare to find networks where a single user access interface or a single transmission technology can cover all requirements in an economic manner. Furthermore, gradual migration means that different stages of network evolution (and hence different technologies) have to coexist inside the network providing full connectivity during the long transition period. In reality, transition is the permanent state of most networks, due to constant arrival of new requirements and new technologies. The network is therefore a conglomerate of transmission media and technologies providing support for the transport of different aggregated data flows and different application network interfaces. Typical architectural models corresponding to transmission and distribution grids communications are presented in Fig. 23.4.

The transmission grid communication network model presents legacy and Ethernet (L2) and IP (L3) services delivered over optical fibers, broadband wireless links and narrowband media such as Power Line Carrier and Industrial UHF radio. The different application network services can be delivered natively over the mentioned transmission media (e.g., teleprotection over PLC, IP, or Ethernet access through optically connected switches and routers, or Current Differential Protection over dedicated fibers. The application network services can also be aggregated over a primary access multiplexer, over an aggregation switch or an SDH/SONET system. The aggregated data stream can itself be transported in different manners, natively or over another layer of telecom technology as represented in the figure. It can be observed that dynamic L3 (IP) routing in this example is limited an end-to-end role (Application Networks) and not used in the Transport Network. A static packet-switched network technology with centralized management (SDH-like) is adopted in particular where abundant legacy services are to be provided (e.g., transmission grid) and wherever traffic needs to be engineered without excessive complexity and in continuity of existing systems. As discussed earlier in this part, MPLS-TP is one such solution allowing multiple levels of logical and

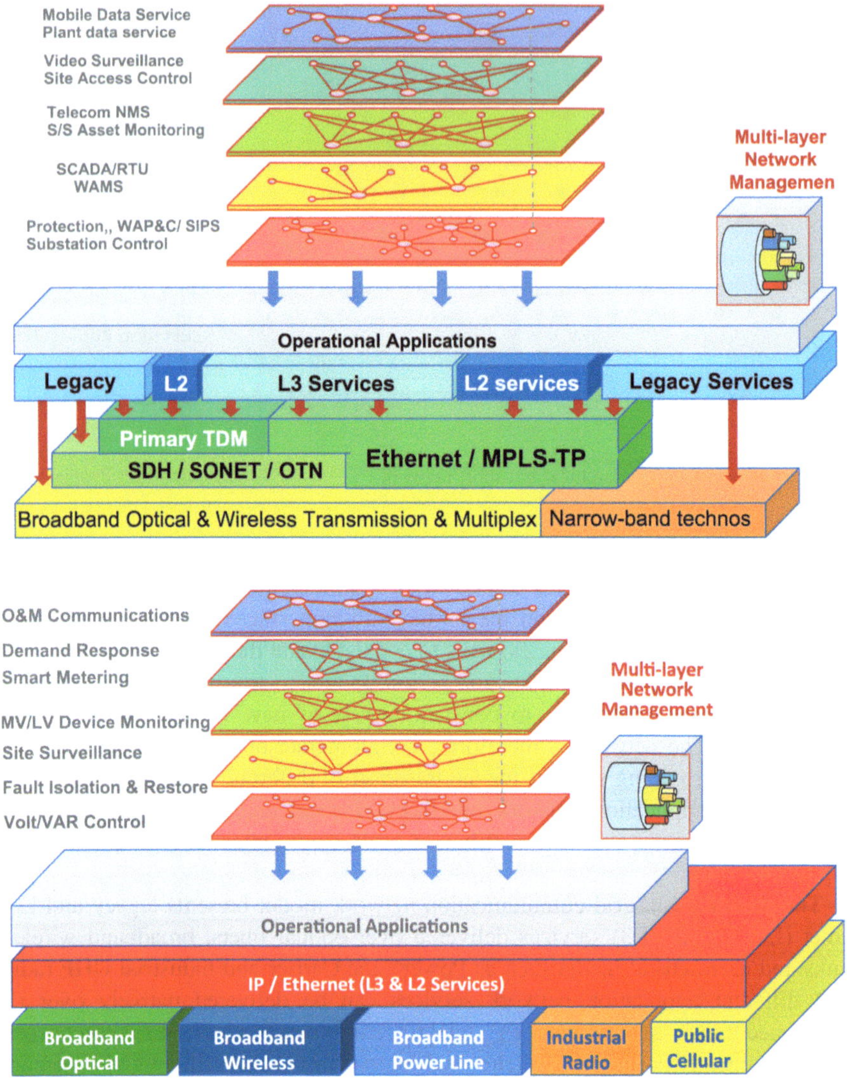

Fig. 23.4 Typical power utility operational telecommunication architectures

physical tunnels as previously deployed by power utilities in their SDH/SONET networks. In this manner, there is no fundamental difference between managing an SDH network and managing a packet or TDM/Packet hybrid network.

The distribution grid communication network model follows the same general philosophy but is composed of substantially larger number of sites, a dynamic network topology, and practically no (or little) legacy. Time constraints are less critical and generally do not include extensive SDH/SONET in their existing infrastructure. On the other hand, covering the distribution grid generally requires the association of multiple transmission technologies depending on urban or rural environment, distances to cover, public network coverage and required bandwidth. IP resilience and Ethernet broadcast are often employed across the transport network to assure sites' coverage providing IP/Ethernet access services to different applications. Broadband media are currently used for Backhauling of services to data processing platforms (grid automation, outage management, demand response, metering, etc.).

24 Revisiting the Process Model—Upstream Management

Electrical power utilities have undergone significant organization change, largely influenced and driven by political and legislative policy, generally moving away from the vertically integrated, government-owned monopolistic organization responsible for the generation, transmission, distribution, and supply of electricity towards a competitive electricity market. This trend has significant impact on the strategic decisions as to the way to provision telecom services:

- The EPU is a commercial organization whose goal may not only be delivering reliable and secure electricity but generating profit through "diversified" activities and paying dividends.
- The stake holders are no longer limited to the state but also the parent company, the investors, the customers, and the regulation authorities.

In such a context, the tasks of Upstream Management for Telecom Services widen up considerably. These cover all the processes resulting in the adoption of a particular telecom service delivery mode and the preparation of financing, facilities, organization, resources, and partnerships for the activity to be sustainable, economically viable, and approved by the different stakeholders.

Referring back to the Utility Telecom Operation Map presented in Sect. 17, Fig. 17.1, an upstream management block was introduced briefly as shown in Fig. 24.1 hereafter.

Upstream management comprises all tasks relative to building or adjusting strategies for delivering telecom services (e.g., procuring services, partnerships, provision and delivery contracts, etc.), necessary material or immaterial capabilities (network infrastructure, tools, processes, skills), and the services to be delivered by the telecom entity. The upstream management prepares an organization with staff, processes, assets, and service contracts allowing the delivery of telecom services within the electrical power utility. It also comprises the migration process for the transformation of service delivery scheme, service, network, or organization to meet new requirements or to improve the quality or the cost of the delivered service.

© Springer International Publishing Switzerland 2017

C. Samitier (ed.), *Utility Communication Networks and Services*,
CIGRE Green Books, DOI 10.1007/978-3-319-40283-3_24

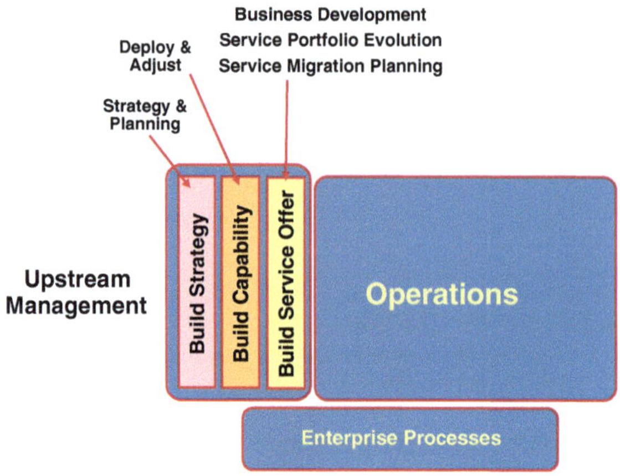

Fig. 24.1 Upstream management in the uTOM

On the other hand, upstream management requires feedback data from the "Current Operations" to compile statistics for the planning of longer term services, network upgrades, tools, and process improvements. Figure 24.2 illustrates the interaction between upstream management and current network operations and Fig. 24.3 the process timeline for upstream and operational telecom network management. A typical process flowchart and organization for these tasks is finally presented in Fig. 24.4 and described in the following paragraphs.

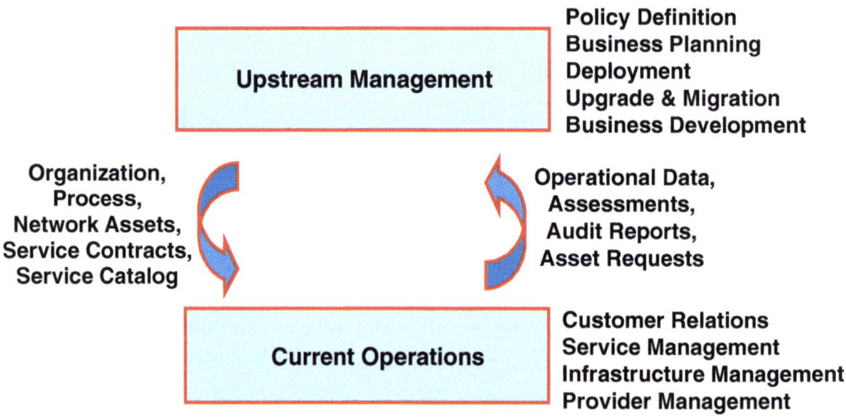

Fig. 24.2 Interaction between current operations and upstream management blocks

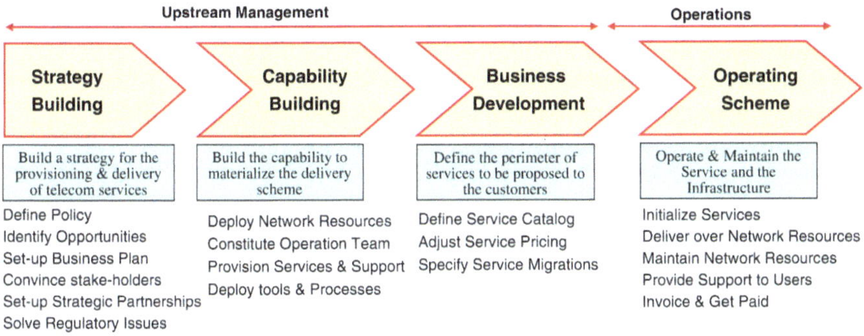

Fig. 24.3 Telecom management process lifecycle

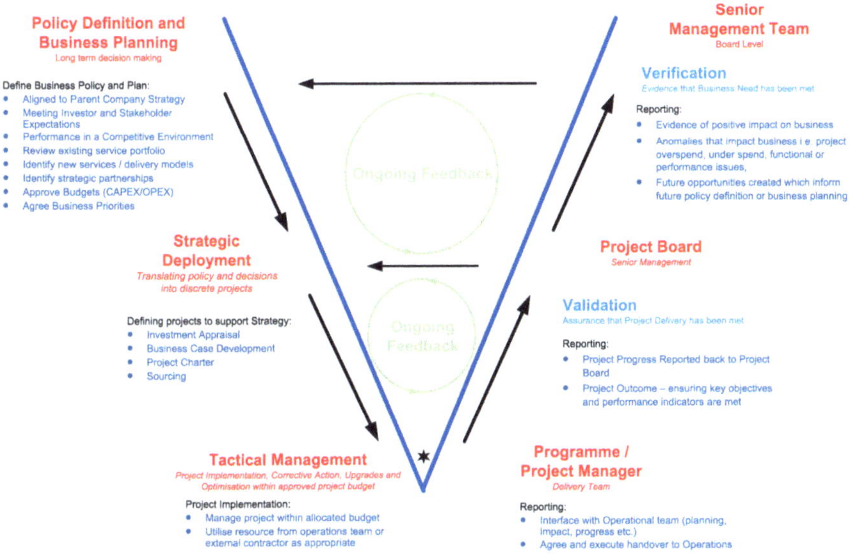

Fig. 24.4 Upstream management process flowchart and organization

24.1 Policy Definition and Business Planning

Telecom policy definition and business planning covers executive level decision making processes that lead to the long-term plan (5–10 years) for the evolution of telecom service provisioning. Its primary aim is to outline a clear direction and framework for how a competitive telecom solution can be provided within the operating environment. The strategic policy on telecom services must identify the focus and strategic direction of the business over the next five year period:

- Determine how the organization will deliver the services based on its existing resources,
- Define any requirement for the cessation or diversification of existing services; identify opportunities or requirements for the creation of new services,
- Quantify any subsequent investment necessary to fulfill such objectives.

The strategic policy for telecom services depends upon a number of factors and information from a variety of sources. Some important factors and issues are enumerated hereafter

Regulatory Context—Telecom activities of the EPU must respond to the requirements of any Regulatory and Legislative frameworks to which it is bound. Regulatory constraints may have significant impact on the availability of funds to develop the telecom facilities, may create new security or safety obligations leading to extended telecom services or refurbishments and renewals, may favor certain delivery modes to others through CAPEX/OPEX distinctions, and may present an obstacle to the integration of corporate enterprise services with the operational services.
Organizational Opportunities—ICT Convergence, the technological trend removing the boundaries between information processing/storage and exchange and the similarity in the required skills is driving utilities to consider the merger of these organizations bearing in mind that the "cost and effort saving" is to be balanced against potential issues regarding certain processes and work constraints. Such a merger between IT and telecom activities can greatly impact the strategic orientations for telecom service delivery. Similarly, the cost and effort for maintaining two independent telecom organizations within the utility perimeter to fulfill operational (OT) and corporate enterprise (IT) communications is driving many utilities to consider their merger.
Market Opportunities—Revenue Generating Commercial telecom involvement through U-Telco business are often envisaged by EPUs every time that major enhancements are planned in the telecom infrastructure. Setting up a U-Telco business, however, necessitates the preparation of a distinct business case based on detailed potential market surveys and opportunity identification which is beyond the scope of the present document. Combining the essentially "non-profit mission" of operational telecom service provision with the "profit-oriented" commercial service in a same organization may rise a number of organizational and regulatory issues that need to be carefully analyzed (e.g., business development in a competitive environment without compromising the critical service constraints of the power system, or access to financing for operational service development while maintaining competitive commercial activity).
Strategic Partnership Opportunities—Many utility services require coverage, workforce, or investment which is economically unfeasible for the EPU on its own. Resource sharing (cable infrastructure, radio coverage infrastructure, emergency mobile network, or maintenance team) between multiple utilities or multiple critical service users can be a way around this economical obstacle.

Investment Context—Financing the deployment of a large telecom infrastructure can be performed through the company's own funds, or through international loans. This latter type of financing cannot be applied to an essentially OPEX mode of telecom service delivery (e.g., procuring telecom services). Similarly, obtaining financial support for developing the network for competitive commercial services requires demonstrated "Return on Investment," which may not be required for financing operational service development. Combining the two may cause difficulties for obtaining either type of financing, unless clearly separated.

Resource Context—As discussed in the next section, asset ownership, in particular optical fibers across transmission lines, is a determining factor on the selection of service provisioning modes and on cost assessment. It is also a determining factor for the feasibility and interest of U-Telco involvement and hence of the envisaged strategic orientations.

Another highly decisive resource in the EPU telecom environment is RF Spectrum. Assuring that adequate RF spectrum is allocated (or maintained) for the operational usage of utilities is a major field of national and international action in many parts of the world in particular in the US and Australia. Gaining (or keeping) access to specific parts of the spectrum is a political issue necessitating power utilities to voice their concerns collectively to legislators and regulators (e.g., UTC). Access (or lack of access) to adequate RF spectrum highly impacts the mode of delivery for mobile services and power distribution network communications.

Skills and Human Resource Context—Unavailability of sufficient workforce and/or skills for scaling up the telecom activity may lead to the merger of IT and corporate enterprise services with the operational telecommunications, or to outsourcing/contracting.

24.2 Strategic Deployment and Tactical Adjustments

These tasks which have been collectively named "Building Capability" transform strategic decisions on the provisioning of EPU telecommunication services into deployed solutions. They comprise a great amount of project and contract management. Strategic deployment may cover all or part of the following processes depending upon the utility organization, regulatory context and the scope, scale, and type of service delivery which is envisaged:

Building a Business Case—Building a business case is at the frontier between strategic decision making and deployment. It transforms the business plan into a detailed project usable for investment appraisal and obtaining required approvals.

The business case in general includes the benefits of the proposed project for the EPU, the estimated cost, the analysis of "Return on Investment" and the risks related to the project. In utility telecom projects, it may also include the risks associated to "not deploying" the proposed project as well as the assessment of alternatives.

Building Organizational Capability—Telecom workforce exists to some extent in all EPUs. It may form a specific organization or incorporated into other operational entities. Organizational changes related to strategic deployments are often the result of major changes in the scope and scale of the delivered services. This may comprise technological change for which the workforce is not skilled and cultural changes needing time to be assimilated (e.g., formal relationships). Change in organization necessitates prior preparation of formal operational management processes as described in part III and further discussed in part V.

The organizational change may be the change of service provisioning mode from procured telecom services to an EPU-operated dedicated network or vice versa. It can also be the change from an operations-incorporated activity into a multi-user internal service provider or affiliated service company as described in part II on service provisioning models.
Building organizational capability may require new skills and more staff than existing in the present organization. This issue can be solved through:

- Employing workforce with required skills
- Extensive technical training programs
- Outsourcing skilled workforce
- Contracting certain support activities (e.g., maintenance)

Deploying Network Infrastructures—Rehabilitation projects leading to major change of scale and scope of the EPU dedicated telecom network are often contracted in a turn-key manner. Contracting of these projects often comprising design optimization, procurement, installation and commissioning of all necessary equipment, provision of power supplies, survey of existing facilities and integration with the existing, necessitates precise specification covering responsibilities and liabilities.

Telecom network deployment projects are rarely "greenfield installations," rehabilitation projects and their network-wide surveys are often the opportunity to review the lifecycle issues of existing equipment: compare the costs of replacement of old equipment with the cost of upgrade and maintain. Timely replacement of old telecom assets may allow substantial economy on operating expenditure (spare parts, obsolescence management, site interventions and repairs, etc.) in addition to enhanced functionalities and performance.
A major issue in large-scale telecom network rehabilitation projects is the migration of operational services from the existing network to the target network with minimal and scheduled service interruptions.

Building Service Contracts—Contracting of services for the deployment of network infrastructure and management tools, and/or operation, maintenance and support necessitates precise specification. If turn-key implementation projects are often well specified, precise, and covering responsibilities and liabilities of each party, service contracts are often poorly defined. Service contracts are often based upon Service Level Agreements (SLA) describing the engagements of the contractor to intervene upon encountered anomalies in constrained time: this contractual time defines the grade of the SLA (e.g., gold, premium, diamond, etc.). The contractor must provide people, organization, knowhow, tools and process. However, the responsibility of the contractor, who has not designed the system, has not chosen and procured the equipment, and has not decided to maintain rather than to replace old assets, cannot be extended to the proper operation of the system with an acceptably low down-time. He can just guarantee to intervene in contracted time limit with skilled staff. Coordinating the contracts for equipment procurement, network implementation and operation and maintenance can prove to be extremely hard, leaving gaps in the overall responsibility.

The liability of the contractor is another important issue. In general, the level of sanction cannot cover the potential loss: risk sharing cannot be back-to-back.
Building organizational capability may also result in the need to outplace existing staff previously involved with that particular service into the contractor's workforce. This delicate human resource issue needs to be taken into account and treated at the upstream management level prior to the change. Moreover, it should be noted that the use of external resources in whichever scheme is indeed a way to acquire rapidly and flexibly the required skills for introducing new technologies, but the obtained skill and experience is not sustained in the company. Outplacement of skills and experience into external contractor companies is often an irreversible process and may represent a loss of control and capability in telecommunications, rendering the EPU dependent of its contractor.

Deploying Management Tools and Processes—Building organizational capability requires also the definition and application of new operational processes which will be appropriate for the new scope and scale of activity. These are described in part V on "Maintaining Network Operation."

Deploying management tools, beyond the vendor-specific equipment configuration and management platforms, represents an investment that many utilities are at present finding necessary. These tools consist of online and offline information systems allowing interactions with the network infrastructure, with service users, with service providers and across the management organization. They include service desks, network and service configuration management data base, alarm and event management platforms, performance monitoring systems, incident management and work order dispatch systems, dynamic route tracking systems, security management and intrusion detection systems, etc. These IT platforms and applications were described in Part III (service management) and further in part V on "Maintaining Network Operation."

Management tools employed in small and simple telecom networks are often trivial, "home-made" and may have marginal cost of maintenance. However, scaling up the network and formal processes necessitate more elaborate tools with higher cost and effort to deploy and to maintain.

Deployment Phasing and Migration plan—An important task in transforming business plans and policies into deployed solutions is the elaboration of a phased deployment plan describing the step-by-step migration from the existing situation to the target solution. The deployment plan may in this way be extended over some years depending upon a number of parameters:

- Business plan requirements,
- User requirements anticipated in the Service Catalogue/Roadmap (see next section),
- Lifecycle of existing telecom assets on the network,
- Deployment capability and skills
- Investment plan and availability of funds,
- Utility's Power delivery mission constraints and minimal disturbance planning.

Validation and Feedback—The process of deploying a telecom solution allows the practical validation of the long-term policy and its related strategic decisions. Parameters and factors which have not been taken into account and hypotheses which turn out to be invalid are in this manner identified. The feedback can be used to adjust the business plan and the strategic orientations. The deployment process must use previously defined metrics (KPIs) and devise measurement capability to validate strategic decisions.

On the other hand, operational management teams can provide, through their processes described further, valuable feedback used for identifying:

- Asset usage and potential optimization
- Asset "replace or maintain" requirements based on operating costs
- Existing service delivery costs
- Service contractors and supplier performance,
- Security reinforcement requirements, etc.

Tactical Adjustment—EPU communication service requirements are not static; they evolve in time. Anticipating new service requirements is described in the next subsection on Service Offer, however, adjusting the capabilities of the network (or provisioning contract) in terms of coverage, bandwidth and network in order to meet encountered requirements is to be treated with a higher reactivity. System and process upgrades and optimizations are performed within the perimeter of the approved yearly budget. This is generally carried out through setting projects and following their deployment using the operational management or contractor workforce. The scope may cover optimization, corrective action, upgrade/renewal of equipment, and firmware upgrade.

24.3 Business Development, Service Offer, and Service Migrations

If policy definition and business planning were assimilated to the decision to set up a restaurant at a certain location, assessing the catering mode (e.g., served at table, self-service, fast-food, etc.) and looking after legal issues, then strategic deployment would be employing staff, purchasing equipment and preparing the premises accordingly. In this case, this section would be about the content of the menu, attributing prices to different items and adjusting the menu in time to correspond to customer tastes and preferences.

In the Telecom Operator world, this is called Telecom product marketing and lifecycle management and is mainly related to market survey.

In the EPU's operational telecom context, building the Service Offer corresponds to the analysis of user application requirements, grouping of constraints and attributes into different categories of Service Level Agreements (SLAs) and hence to build a "Service Catalogue."

Building the telecom service catalogue also comprises the determination of service prices based upon the initial network investments (cost of building the capability) and operating costs (operational management cost), this latter also including all external costs due to suppliers and contractors. The manner in which these costs are converted into service prices are governed by the profit strategy set up by the Business Plan (Strategy Building).

Pricing of services may be used for direct invoicing of services to the users or more often for the repartition of expenses:

- between internal and external users
- between operational and corporate users
- between different operational entities

As for the restaurant menu, the communication service catalogue needs to be updated according to encountered and anticipated changes in the EPU's applications. Typically, SCADA applications are in most EPUs migrating from serial point-to-point communication channels to Ethernet-based TCP/IP networking. This change, together with other applications' migration to Ethernet, is reducing considerably the requirement for serial point-to-point communications, but increasing sharply the requirement for time-controlled wide area Ethernet. The service catalogue must therefore be adjusted correspondingly, leading to the deployment of adequate capabilities in terms of network, management tools and skills.

Service migration planning is often to be phased according to a number of different factors:

- Application asset lifecycle and migration plan
- Extent of required infrastructure change and corresponding investment plan
- Deployment capability for the migration and readiness of the organization
- Other planned changes and extensions, allowing a grouped migration project, reducing costs, and service interruptions

Telecom Network Asset Ownership 25

Ownership of major telecom assets is a determining factor in the EPU's adoption of a telecom service delivery model, and on the EPU's degree of control over its operation-related communications.
Telecom assets in the EPU can be broadly classified into the following types:

- Physical layer assets—Optical fibers, right-of-ways, cable trays, RF towers, frequencies.
- Transport network assets—All electronic equipment used for the core transport of information. We purposely separate the assets for the bulk transport of information from those used for multiplexing and interfacing of individual applications which constitutes an edge or a distribution layer, even if in technological reality these two layers can at times be merged together and therefore render difficult the separation of their assets ownership.
- Application service network and platform assets—all specific systems delivering particular communication services through the core transport capacity (e.g., low capacity access multiplexing, voice network, protection signaling, SCADA communication network, etc.).

We classify ownership patterns into three broad categories:

- Assets owned by the service user (the operational entity in the EPU)
- Assets owned by the service provider (whatever be its relationship with the user)
- Assets owned by another party (e.g., state-owned fiber, another provider, another utility, etc.)

This section analyses some specificities in each case. Ownership criteria and issues for some common telecom assets as described in the section are summarized in Fig. 25.1.

© Springer International Publishing Switzerland 2017

C. Samitier (ed.), *Utility Communication Networks and Services*,
CIGRE Green Books, DOI 10.1007/978-3-319-40283-3_25

Asset Layer	Asset Types	Ownership Criteria & Issues
Physical Layer Assets	Optical Fiber Conduits, Rights-of-way RF Towers, Repeater Housing Radio Spectrum and Licenses Long Term Contracts	HV Transmission Lines Civil Works & Access Rights Suitable Premises Regulatory Constraints Legal Constraints
Transport Layer Assets	Bulk Data Transfer Connections Core Network Infrastructure Narrowband Telecom Links (PLC, Radio, etc.)	Cost of Ownership Bandwidth Requirement Lifecycle Issues & Upgrades Required Level of Control Availability of Expertise & Skills
Application Layer Assets	Service Multiplexing Teleprotection Signaling Voice and Data Servers LAN/WAN Assets Surveillance Systems, Platforms	Critical Applications Coupling Cost of Ownership IT Lifecycle Issues & Upgrades Availability of Expertise & Skills

Fig. 25.1 Utilities telecom asset ownership

25.1 Fiber and RF Infrastructure

Fiber, Optical Cable, Right-of-way

The most determining physical asset in the utility telecommunication network deployment is the optical fiber. It can lead an EPU to a particular mode of service provisioning or prevent an EPU from adopting a particular mode.

Installing optical fiber cables between communication sites of the EPU necessitates "Right-of-way," that is to say underground or overhead corridors where the cable can be laid. This is a very precious asset that transmission utilities own due to their HV transmission lines.

Optical fiber infrastructure can be provisioned by the EPU through one of the following manners:

1. **Procure and install fiber cables**—through the EPU's right-of-way corridors (overhead power lines, underground power cables, etc.). This is by far the most used scheme in Transmission Utilities, and in Distribution Utilities owning HV lines. Spare capacity can be used for corporate and other communication services and spare fibers can be leased to external users for covering costs or for extra revenue. Possession of extra fibers may lead the EPU into building a U-Telco activity.
2. **Jointly financed procure and install**—This scheme is typically employed at the interconnection between two EPUs, e.g., transmission line interconnecting two transmission utilities

3. **Fiber (or service) in exchange of right-of-way**—The EPU fiber requirement being far lower than the capacity of an OPGW, it can grant a telecom carrier the right to draw multi-fiber OPGW cables in exchange of its required fibers in those cables. However, this scheme presents many issues concerning the maintenance of the OPGW which is intimately related to the maintenance of the transmission line. Even if often envisaged (e.g., for immediate availability of financing when a sizable fiber infrastructure is needed), it often evolves into case 1 with leasing of extra capacity. However, where the telecom entity of the EPU moves away into commercial service and becomes a distinct company, it may inherit the fibers and consequently the EPU's right-of-way through a long-term leasing contract, in exchange of fibers or services left to the EPU.
4. **Swap with other fiber asset owner**—This is typically used for providing route redundancy where the network's topology does not provide the required resilience. The other asset owner can be another utility, a telecom carrier, etc. Access from the fiber asset owner to the EPU site may be an important issue. It should be noted that these swapping schemes may raise regulatory issues regarding the nonpayment of taxes and duties.
5. **Lease fibers in another EPU's cables**—This scheme is often employed at sites where a smaller footprint EPU connects to a more extended EPU. Some typical examples are:

   - generation plant using transmission utility fibers at the transmission grid substation,
   - distribution utility access to a national facility using transmission utility fibers.

   It can also be used to close a partially open telecommunication ring using assets belonging to a regional footprint utility. Dark fibers (rather than transmission links) are leased when the distances are sufficiently short to avoid intermediate regeneration and when high capacity is required (e.g., Giga Ethernet). Fiber leasing from another EPU is generally performed at co-located sites and therefore avoids the "last mile" issue encountered in other leasing schemes.
6. **Lease from a fiber asset owner**—This is the typical situation for EPUs that require a high degree of control over their telecommunication network but do not have the necessary right-of-way (or the initial capital investment) for installing their own fibers.

It should be noted that optical cables over transmission lines may also be under state's public property conceded to the TSO for internal usage. This type of long-term concession in general does not authorize the entity to which the cable is conceded to lease dark fibers.

Using leased fiber from an asset owner other than another EPU, raises several important issues that need to be considered:

(a) The topology of the resulting physical network depends upon the fiber owners' network structures leading to far longer than necessary links and often far from optimal for the utility.
(b) "Last Mile" issue—The distance from the fiber asset owner access point to the EPU premises, even if relatively short, needs right-of-way and non-negligible civil works inducing important cost and delay consequences.
(c) Physical routing issue—The design of a fault tolerant transmission network is based on the knowledge of physical medium routing which is controlled by the fiber asset owner. In particular, where fault tolerance is concerned, the two routes must not pass into any common nodes, cables or conduits. In the event of "incorrect" routing information or cable route changes, the fault tolerance of the whole system may be unacceptably compromised. The EPU has no other way than the "provider's word" to keep track of changes or of the correctness of the routing information. Moreover, it is particularly hard to obtain two independent cable routes from the EPU premises to the cable provider's "meshed network" (which may not coincide with the provider's access point).
(d) Maintenance works—EPU need to have control of maintenance schedules which is not the same thing as being informed of the date of maintenance works. A multi-customer fiber provider cannot program his works according to EPU requirements. In case of interruption of utility leased fibers, EPU requires immediate repair which leading to un-scheduled interruption of other fiber users without prior notice. However, in case of other users' fiber interruption, the EPU cannot accept non-anticipated maintenance works. A very nonsymmetrical contract, in general unacceptable to the provider, is needed by the EPU.
(e) Cable reliability—The majority of fiber providers have underground cable infrastructures, particularly subject to cable cuts due to civil works, while overhead OPGW normally used by the EPU is almost invulnerable. The extremely high levels of service availability required by EPU operation-related applications are very difficult to meet with the probability of cable cut that can be obtained from cable providers. (For OPGW, cable availability can be neglected in comparison to equipment availability; with leased underground cable the reverse situation is to be assumed).
(f) Multiplying the cable providers in order to meet the necessary coverage of the EPU shall multiply the formerly mentioned issues and creates additionally an important issue of governance and contract management with several contractors and sub-contractors in some cases along one same connection.

RF Physical Layer Assets

Another important category of physical layer assets in the EPU are those related to implementing radio transmission networks and links. We add repeater housings under RF assets even if it applies also to optical regenerator housings.

RF Towers in HV substation premises and power plant sites, including tower lighting and its associated power supply, are generally the property of the EPU. These towers as well as other EPU structures which can serve as antenna support (e.g., electrical poles, power plant tower structures, etc.) may also be used by other radio network infrastructure owners such as cellular radio operators as a source of extra revenue for the EPU. Antenna support outside EPU premises (e.g., microwave repeaters or radio base stations) can be through co-location on towers belonging to other radio networks. In particular when a wide zone coverage is required (e.g., UHF data systems, mobile trunk systems, etc.), the optimal location of radio relays for covering a given zone is often the same for all radio infrastructures, facilitating co-location.

Repeater housing, including air-conditioning facilities and repeater power supply can be the property of the EPU, its telecom Service Provider, or leased from an external party. Microwave link repeaters are often located on EPU premises in which case, they are generally EPU assets. UHF and other zone coverage base stations on the other hand, are often on high sites, and may be in shared housing leased from another asset owner or telecom service provider. The maintenance of the facilities, in this case, is generally provided by the owner as an external service to the EPU. When using externally provided power supply for radio relays, the autonomy of the power supply and the dimensioning of batteries are important issues for operation during power outages.

Frequency licenses with narrow directivity (e.g., microwave radio links) or narrow bandwidth (industrial UHF radio for SCADA) are generally obtained by the EPU and are therefore part of its assets. Licensed broadband spectrum with wide coverage, on the other hand, cannot be allocated in many countries for the exclusive usage of the EPU internal communications (e.g., Cellular systems such as LTE). It is, in this case, common to obtain shared usage with other Utilities (e.g., gas, water, other EPU) or other critical services. This is normally performed through procurement of services from a specialized operator, or setting up a service that can be procured by other users. This latter case generally results in the separation of the Service Provider entity from the EPU operational entity.

25.2 Transport Network Assets

Transport layer assets are those related to the bulk transport of information. Optical and microwave radio communication equipment and core data network infrastructure constitute the basic elements of this layer.

If the ownership of physical assets is often a determining factor on the telecom delivery scheme, transport layer assets can much more easily be procured by the EPU if it intends to own its assets.

Asset ownership model at this layer is often based on the following factors:

- Ownership of underlying physical layer assets—When the physical layer assets are not under EPU's control, it is easier to admit lack of control over the transport layer (e.g., leasing STM1, E1/T1 or Giga Ethernet connection rather than leasing dark fiber and repeater housing, power and maintenance). This may lead to more straightforward contract and SLA management and less interactions.
- Required communication bandwidth—Narrowband information transport on owned physical assets, e.g., HV PLC access to substations or wireless SCADA systems, is always performed with EPU-owned transport network assets. On the other hand, when the communication requirement are small compared to the capabilities of the available or suitable communication technology, bandwidth sharing with other users is necessary, either to justify cost, or to overcome regulatory constraints. (e.g., Broadband wireless data services, Satellite Communication Hub).
- EPU's required level of control over the service—The more a communication service is critical in the EPU's process, the more it is inclined to keep full control of the associated transport layer assets (e.g., communication services for Protection Relaying applications).
- Total Cost of Ownership, Asset Life Cycle, and Return on Investment (ROI)—The cost of implementing and maintaining a particular type of transport asset may lead the EPU to renounce to its ownership. This indeed is to be traded off with the requirement to keep full control. It should be added that unlike physical assets, transport network layer assets have much more limited life cycle, meaning that the ROI must be possible in shorter time.
- Required skills for managing and maintaining the transport assets—The EPU may simply not have the necessary skills, tools, and organization to run a particular type of transport network, or the organizational capability to keep it up-to-date. Large core data networks and operator-grade network operation centre facilities are typical examples of "hard-to-maintain" assets.

25.3 Application Service Networks and Platforms

Application service platforms as those necessary for utility switched voice and data services generally belong to the EPU or its telecom service provider. Those which are more intimately related to the operational process or to the operational sites such as teleprotection signaling or SCADA communications are the property of the EPU operations, while those that are shared between operational and non-operational activities (e.g., computer networking, voice, etc.) are often procured and renewed by the Service Provider entity. The short lifecycle of the assets (e.g., IT platforms) and the total cost of ownership being mainly driven by the cost of upgrading and maintaining, the EPU may be inclined to procure services rather than assets.

Planning Network Transformations and Migrations 26

Before undertaking major network transformations and deploying new technologies it is essential to define a consistent architectural evolution strategy for the communication network. The evolution strategy takes into account short, medium and longer term requirements, expected time-lines for new operational services, as well as existing infrastructures, network operation processes and present organizational capabilities. A network design over-dimensioned for present and near future requirements may result in excessive investments for unused capabilities: long term requirements and market-available technologies will both evolve in time rendering the seemingly "future-aware" network unsuitable or non-cost-effective in the future. In the meantime, the over-dimensioned design will have absorbed enormous operational effort and caused numerous network outages due to inadequate skills and processes.

A network transformation plan must provide a migration roadmap through a time-phased sequence of network modifications and upgrades with phased investments and gradual build-up of skills, process, and tools to deliver required capability at each step according to application requirements and to adapt and adjust according to new requirements and new enabling technologies.

Many power utilities are planning to introduce or to extend packet-switching capabilities in their essentially SDH operational telecommunication networks. Some of the reasons driving them to packet network deployment are presented in Fig. 26.1.

Packet switching is being implemented in the power utility operational telecom network in a number of manners with substantially different capabilities regarding time-sensitive critical applications. The employed technologies and the architecture depend on the envisaged applications as well as the size and required coverage of the network.

Packet data communication for each application can be implemented through pre-allocated dedicated bandwidth across a shared transmission network. This is presently the case where the nature of data traffic and the "small scale" leave little benefit expectation from data network resource sharing. The absence of resource

© Springer International Publishing Switzerland 2017

C. Samitier (ed.), *Utility Communication Networks and Services*,
CIGRE Green Books, DOI 10.1007/978-3-319-40283-3_26

Fig. 26.1 Utility drivers for packet network deployment

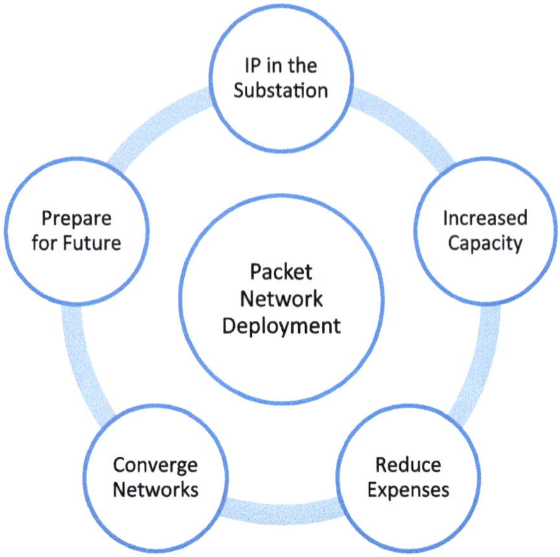

sharing avoids conflicts, priority management, queuing issues, and consequent lack of deterministic behavior.

On the extreme opposite, all packet streams resulting from the transformation of information exchange pipes into data packets may share the same network for a more efficient usage of resources. Multi-service IP integration of all telecom traffic should in theory be the most economical way of implementing data exchange as performed for example in the Internet. However, applied in a straightforward manner, this means non-deterministic and traffic-dependent data delivery.

In between these two extremes, by injecting complexity into the straightforward "shared resource" system we can implement differentiation mechanisms for a better control of priorities, traffic shaping and routing as described in Chap. 22. However, by doing so, we move gradually away from the objective of efficient resource usage. Differentiated services, resource reservation protocols, label-switched forwarding techniques, etc., are steps towards deterministic Quality of Service and steps away from efficient resource usage.

The right mix of reserved/dedicated resources and shared resources, and the compromise between complexity and efficiency depend on many network and traffic parameters, cost and availability considerations, utility workforce, etc.

Figure 26.2 summarizes some main aspects in the balance between dedicated and shared resource network operations.

Operational migration—which scope and which time-scale?

Before going any further, it is worthwhile to note that the necessity for packet networking and the consequent transition from existing TDM to packet-switched technology in the operational network is generally a very slow and gradual change requiring stable and secure operation during the whole process which may take

Dedicated Resource	Shared Resource
• Predictable Behavior – Latency, Routing, Restore	• Equipment Cost Saving and wide spread usage
• Adapted Monitoring and Fault Supervision	• Centralized Monitoring and Fault Supervision
• Adapted Maintenance Planning	• Unified Maintenance Planning
• Adapted Life-cycle Management and Gradual Migration Planning – upgrade or replace only when required	• Flexible creation and modification of connections
• Investment justified through Operational Need for Power system Security – no need to justify by ROI	• Easier to justify new investments through ROI
• Organization & Workforce – Dedicated Staff	• Reduced and Multi-purpose Workforce
• Implicit security through resource isolation and less accessible because of less widespread usage and public familiarity	
• More controlled and easier access to qualified technology	• Wider market availability and higher evolution perspective

Fig. 26.2 Dedicated versus shared resource in utility telecom networks

several years. For a large number of power utilities, the immediate requirement is to provide packet capability to their evolving (or new) substation services and to build "preparedness" for facing new network technologies in future. The most largely deployed applications at present consist in SCADA, IP-based voice and remote management of telecom assets. These IP services require little bandwidth. Tunneled Ethernet over time division multiplexed channels is widely used in substation networks for SCADA-type applications without any substantial change to the network. Ethernet over SDH is described in Sect. 22.6. The systems provide E-line (point-to-point) or E-LAN (multipoint) connectivity with operational grade security and availability, similar to non-switched TDM-transported services. Strong service isolation from other concurrent telecom network users, minimal and deterministic delay, short service restoration time after failure and a high Quality of Service due to dedicated resources are some characteristics of this solution.

On the other hand, transporting legacy TDM over a purely packet-switched infrastructure, although feasible, is much more challenging and may end up being constraining, costly and with moderate quality. It may provide a solution for dealing with marginal legacy in a "mostly packet" service mix. Fully replacing TDM by packet switching in a mainly legacy interface substation is not a viable solution.

Figure 26.3 presents a migration path from TDM to Packet passing through Packet over TDM, Hybrid TDM/Packet (e.g. 2 Networks with or without common resources), and TDM over Packet.

Packet-Switching Beside SDH

For packet switched capacity beyond that offered by the present day Ethernet over SDH, distinct fibers or wavelengths can be used for building native Ring-protected Ethernet or MPLS. In most utilities extra fibers are available in optical cables and the multiplicative potential of wavelength multiplexing is barely used. Adequate optical line interfaces allow a line capacity of 1–10 Gbps without much effort.

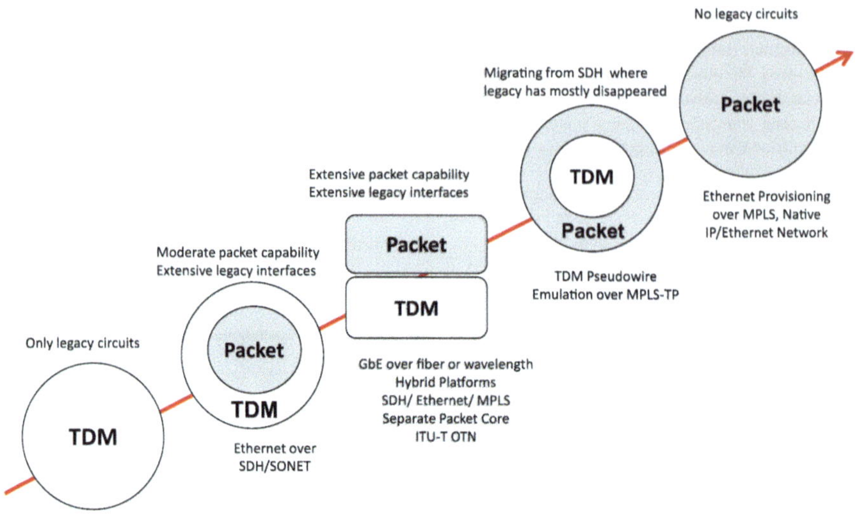

Fig. 26.3 Migration path from TDM to packet in the power utility operational network

In this architecture, legacy services may be maintained over the existing network and gradually migrated when replaced by an IP-based alternative. This is for example the case for SCADA RTUs gradually migrating from serial to TCP/IP. This architecture is presented in Fig. 26.4 hereafter.

Connecting Ethernet switches on separate fiber or wavelength can be an evolution from Ethernet over SDH transforming the existing system into a Gigabit Ethernet in parallel to the existing SDH. The SDH system can still transport low bandwidth Ethernet in its payload for access into substations without native

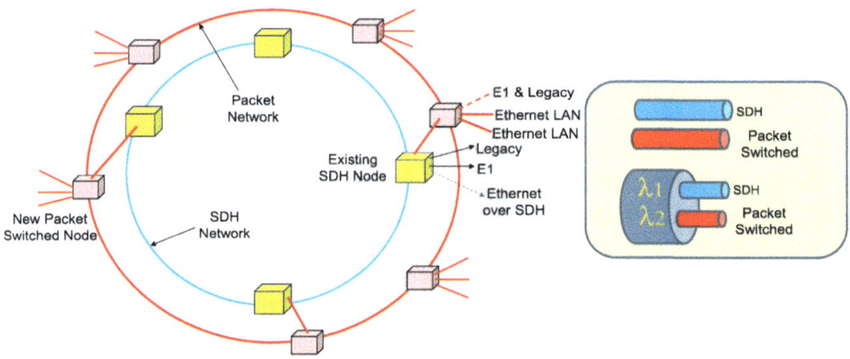

Fig. 26.4 Ethernet or MPLS beside SDH over separate fiber or wavelength

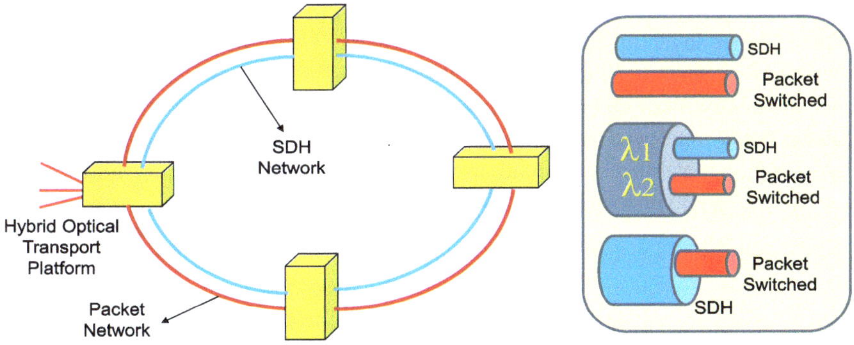

Fig. 26.5 Ethernet or MPLS-TP and SDH in a hybrid platform

Ethernet. For substation applications, extensive packet capacity can be added to the existing system through this simple operation.

The existing SDH infrastructure can, moreover, be replaced by a "hybrid" platform providing both native Ethernet and SDH interfaces in the same equipment hence replacing the SDH and the Ethernet switch as presented in Fig. 26.5. A hybrid platform provides a conventional TDM replacement for the legacy interfaces while adding new packet capabilities for new services. Distinct line interfaces can be employed for TDM and packet communications. These line interfaces are connected to separate fibers or multiplexed over distinct wavelengths on the same fiber.

Building Scalability into the Data Service

Scalability in the utility telecom network is the facility to evolve to a much larger network, larger data capacity and larger number of transported data connections without resulting in extremely laborious service configuration or inadequate quality of communication service.

Ethernet path protection is performed through different variants of spanning tree protocol, or ring protection none of which is suitable for large networks. The Quality of Service in Ethernet is based on simple priority assignment consisting in the allocation of a priority level to each traffic and different queues at each switching node with no guaranteed (committed) bitrate for any data stream.

As already stipulated earlier, however, building scalability needs further knowledge of requirements. It is necessary to examine the required level of scalability, the time-scale associated to it, as well as the traffic exchange requirements across the network. Very often, the intensive traffic growth is not at the substation access but in the transport core, central platforms, and in the office networks.

The separation of aggregation and access network from the backhaul network (Network Core) and the consequent hierarchical architecture is an appropriate way to introduce a gradual change of scale in the utility network. The core network can be deployed without disrupting existing connections which may gradually migrate to it.

Alternatively, the existing SDH infrastructure can be replaced or supplemented with a higher layer of optical transport network (ITU-T G.709 OTN) described previously in Sect. 22.5. OTN allows the transport of existing SDH with Ethernet data streams over a higher bitrate line interface (e.g., 10 Gbps for OTU2). It is at present being employed in a number of electrical power utilities to provide a very high level of scalability at the core level of the network. It can be assimilated to a scaled-up alternative for Hybrid TDM/Packet platforms.

27 Cyber-Secure and Disaster-Resistant Communications

The term "Security" is used with different significations in the power system context:

- Security of the power system is related to the "ability of the bulk power system to **withstand unexpected disturbances** such as short circuits and non-anticipated loss of system elements due to natural causes". Power system security relates to protection and control, remedial actions and other power applications briefly described in part 1. Communication is an enabler for these applications (e.g., Protection Relaying).
- Security as the reliability of communications put in place to serve the power system is related to the **proper transfer of information** across the power system in adverse conditions, in case of infrastructure faults and failures, avalanche of traffic, human errors and malicious interactions, assuring hence the proper operation of different power system automation applications and processes.
- Security as the **protection of information**, relating to utility customers' personal data and to the electric power system, that may be of interest to malicious parties who wish to harm or earn profit from the utility or its customers

Security, as discussed in the present section, concerns issues of malicious or accidental interaction in the operational communication network, and the ways this risk can be minimized with reasonable cost and effort.

Security can be characterized through the levels of Confidentiality, Integrity and Availability that an information-related system must attain.

© Springer International Publishing Switzerland 2017

C. Samitier (ed.), *Utility Communication Networks and Services*,
CIGRE Green Books, DOI 10.1007/978-3-319-40283-3_27

Confidentiality	• Protect information from disclosure to unauthorized persons and processes
Integrity	• Protect data from being modified or deleted in an unauthorized and undetected manner
Availability	• Maintain systems, information and applications accessible and usable upon demand by an authorized entity

Disaster-resistance or disaster-preparedness, on the other hand is the capability of the system and the delivered services to survive natural catastrophe, accidental or intentional disasters causing considerable damage to the network infrastructure, tools or staff employed for delivering operational communication services.

The two subjects of cyber-security and disaster-proofing have many common points some of which are given below:

- Both cyber-security and disaster-proofing are risk-related issues. A risk is characterized by a likelihood of occurrence (average occurrence frequency) which, by nature is difficult to establish. It is given a value based on past experience which must be corrected periodically.
- Both cyber-security and disaster-proofing are to be evaluated through their potential impact on the service or on the utility, in terms of safety, environment, reputation, power loss, economic, legal liability, etc.
- In most cases, the threat cannot be eliminated but the risk or its impact can be attenuated through specific measures. The idea is to push the risk and its impact to a reasonable level depending upon the extra time, loss of performance or efficiency, and extra cost that the utility can allocate for mitigating the risk.
- The solutions in both cases are partially in the work process and precautions, and partially in the deployed infrastructure, risk barriers, architectural design precautions, etc.
- A cyber-attack can cause a catastrophic situation such as power outage or unavailability of major telecom resources in the network forcing the telecom staff into a disaster recovery operation. A natural or accidental disaster situation can push the network in a non-optimal operation service survival mode which may present serious cyber-vulnerability. The two subjects cannot be explored fully independently.

27.1 Risk and Impact Assessment

Figure 27.1 hereafter presents an example of risk management analysis table providing admissible and inadmissible zones according to impacts and likelihood of risks. The table is composed of two parts: Impacts and Average Occurrence Frequencies.

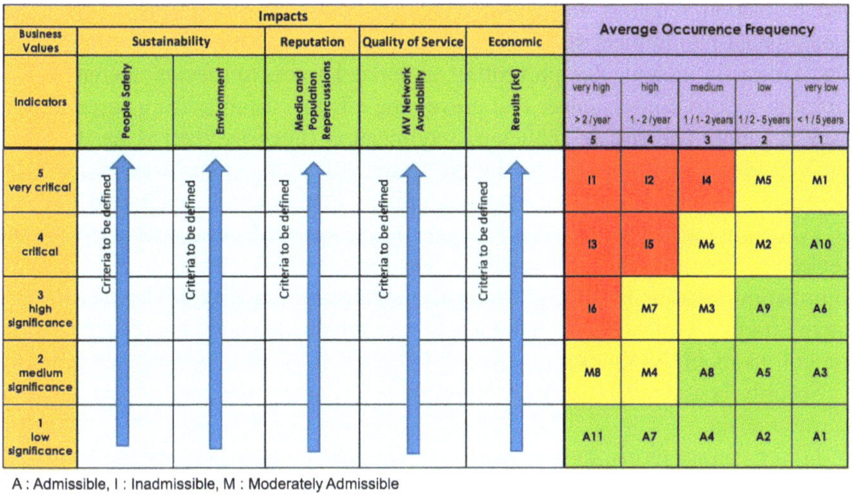

Fig. 27.1 Example of risk management analysis table

A number of potential impact types associated to different risks are identified and in each case scaled with increasing situational gravity from 1 (low significance) to 5 (very critical). The criteria must be defined in such a way to be correctly identified in real situations.

On the Occurrence Frequency portion, the scale is partitioned into different occurrence values ranging from very high to very low. The actual values depend on the Utility. Each intersecting pair "impact level and occurrence" is colored as Admissible, Moderately Admissible or Inadmissible.

The resulting chart is used to analyze different situations and judge upon the necessity to mitigate or not to mitigate any possible risks.

27.2 Designing for Cyber-Security

Cyber-security in the electrical power utility has been, and still is, the subject of numerous studies and investigations with abundant specialized literature beyond the scope of the present book. These studies cover both the power utilities' evolving information infrastructure in general, and the new electrical grid substation automation systems. Here, the objective is to present only some key concepts and issues more particularly related to operational communications.

Designing an adequately secured system is based on risk and impact analysis as described previously. Confidentiality and integrity are the predominant objectives in securing sensible IT systems: it is often preferable that a transaction be blocked rather than disclosed or modified. However, in the case of substation automation and protection relaying, the situation is very different: availability is often the predominant requirement since the system needs to remain operational at all time. One can easily imagine that preventing authorized access to substation controls has far more serious consequences that preventing a bank customer from accessing his bank account. The risk of malicious modification (integrity risk) is on the other hand moderate as long as no public communication network or wireless is employed, although accidental interaction due to human errors (leading to integrity or availability threats) in a complex and interconnected environment cannot be neglected. Similarly, the consequence and impact of unauthorized disclosure of information remains relatively low in the automation context unlike metering or energy trading applications. Nevertheless, one should also note that security threats are not invariable in time.

Security is assured generally at multiple levels across the peer-to-peer communication chain.

In very broad terms, cyber-security in the power system covers the following concepts:

Security Policy and Process—Before being a matter of tools and techniques, cyber-security is related to the processes and practices in the operational organization of the Utility, specifying rules on "who can access to what," defining ways to assure this, and checking continuously if the rules are respected and the goals achieved. Devising a security process necessitates the survey of applications, critical assets, and user profiles.

Electronic Security Perimeter—Services and applications are segregated into different Security Zones based on the consequences of a security breach, with data exchange between these zones undergoing restrictive controls according to the security policy (e.g., firewalls). A zone model can enforce the termination of all inter-zone connections in a DMZ server (Demilitarized Zone), forbidding hence the initiation of direct connections across zones (see Fig. 27.2).

Device Access Control—Local and Remote access to devices (application or network) are controlled through the use of certificates and passwords as well as encryption when traversing unsecured networks. Peer authentication (application level) mechanisms are generally used for remote access.

Network Access Control—Application Network access at the Transport Network's Service Access Point may be controlled through an appropriate security barrier specified by the Application Network user.

Transport Network Service Isolation—The transport network may provide security through encryption and/or low layer isolation of different transport services (VLAN separation, time slots, etc.).

Securing Operational Communications

Telecommunication links accessing the substation carry traffic corresponding to distinct applications (protection and control, SCADA, surveillance, voice, etc.). Partitioning the intelligent electronic infrastructure into separate security zones

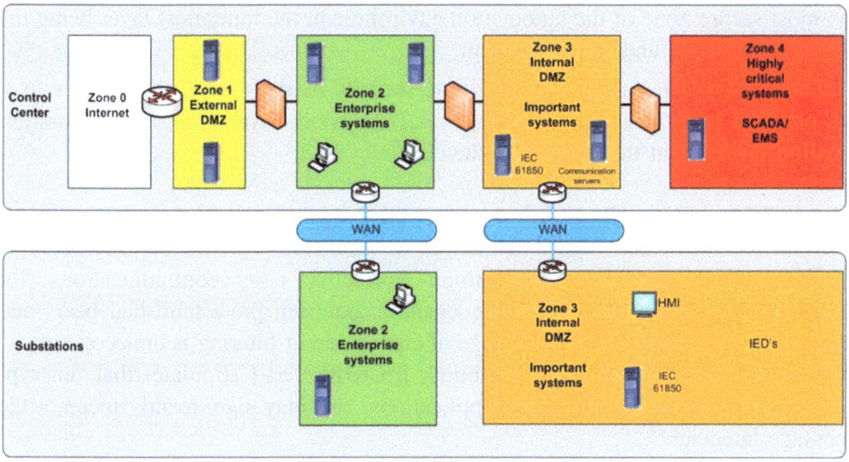

Fig. 27.2 Example of security zones (*IED* Intelligent Electronic Device)

implies that the aggregate communication traffic of the site must also be partitioned into separate networks. One must therefore assure isolation between different networks having access to the substation through the same aggregated telecommunication links. Traffic separation and isolation is performed either through implementing distinct networks each with its dedicated equipment (e.g., separate routers and switches) or by using virtual separation through common network nodes (i.e., VPNs). In practice, a combination of physical and virtual separation is often employed to partition the aggregated data-stream of the telecom connection.

Concerning the wide area network, IEC61850-90-1 stipulates that "If some of the Ethernet Telecommunications network equipment is outside the utility's Security Perimeter (e.g., when Ethernet circuits are leased from a service provider), the links through such equipment should be secured through a technology such as 'L2TP' (Layer 2 Tunneling Protocol) to create a VPN, and the security should be maintained through the implementation of the associated key-management requirements." In other words, as long as one remains on a dedicated fiber-based private network, no further security measures need to be taken on the WAN other than channel isolation. The security barrier is located at the access point of the automation LAN (Electronic Security Perimeter) and then further access control is performed at the protective relay device communication port. Combining multiple layers of security is generally considered as the efficient way to prevent unauthorized access.

LAN/WAN Network Access Control and Isolation

As already described, device access must be protected at all ingress ports, either at device level or at network level or both. In practice a layered security protection approach is desirable in which multiple levels of firewalls and demilitarized zones (DMZ) constitute several barriers from the largest attack surface (most public) to the most secure zone of the automation environment, the innermost layer being the device's password and authentication.

The employed techniques to constitute security barriers and isolation operate at the physical layer (L1), data link layer (L2), network layer (L3), higher communication layers or at the application level.

Physical Layer Security—This concerns the usage of a dedicated physical medium such as dedicated fiber, wavelength or time-slot (e.g., E1 or Ethernet over SDH) for accessing user applications. Protective relay communications, for example, are continuous, very time-sensitive, between pre-established peers and over a closed network (or link). Usage of encryption or filtering is unnecessary and may impact their operational availability. IEC61850-90-1 stipulates that "encryption is not required for protection applications, and may significantly increase the message latencies."

Particular attention should, however, be paid to the proper security of the Telecom Management System from which the configuration of the telecom system and therefore the protection channel can be modified or interrupted for time-slot isolated channels.
Physical separation at the substation local level corresponds to no connection which is an efficient way to assure security when interactions are not needed. Some utilities may have requirement that the operational network is isolated from the service network and remote access to devices in the operational network is only possible from dedicated terminals that lack any physical ports like USB interfaces, CD-ROMs, etc. The only way to copy data to these specialized terminals is via dedicated access management servers.

L2 Security—Link layer mechanisms such as Ethernet VLAN (IEEE 802.1Q) may also be used to isolate both LAN and WAN level information exchange. In particular, when a separate L1 channel cannot be allocated to the wide area connection between protective relays, then a VLAN is the second best solution as already discussed in the report. IEEE 802.1Q provides the ability to logically segregate traffic between predefined ports on switches.

For an L2 Ethernet connection to be secure, it is again essential that the access to Ethernet switches and the associated remote management system are adequately secured. This can be performed through the following mechanisms:

- Multi-level user passwords securing the switch against unauthorized configuration as well as encryption of passwords and other control information crossing the network (SSH/SSL)
- SNMPv3 encrypted management communications
- Unused ports of Ethernet switches must be shut down

- Used ports are to be secured using MAC-based port security (i.e., only predetermined MAC address devices can communicate via the port) or preferably through IEEE 802.1x (Port-Based Network Access Control). IEEE 802.1x is an L2 mechanism allowing the configuration of Ethernet switch ports so that only authorized clients can communicate using them. The user device requests network access from the switch through its 802.1x agent and an Authentication Server receiving RADIUS messages (Remote Authentication Dial in User Service) checks the user's authentication credentials (Line and System Protection using Digital Circuit and Packet Communications 2012).

L3 Security—IP firewalls are generally used to separate Security Zones. Firewalls are go/no-go portals through which all data must pass in order to enter or exit a network. Filtering is performed by comparing the IP address of incoming packets to an Access Control List (ACL).

Stateful Firewalls perform a partial analysis of the content (e.g., Message Headers) and are hence somewhat in-between Packet Filter Firewalls and Application-level Gateways described below.
IP VPN tunnels using IPSec protocol constitute a common way of isolating communications across a multi-service IP network. However, in the context of protective relay access, we are rarely in this situation unless technician access from outside of the utility network is required. In practice, Layer 2 tunneling, Ethernet VLANs and MPLS VPNs are more common.

Higher Layer Network and Application-level Security—Many of the security mechanisms mentioned above even if they operate on data at L2/L3, they make usage of higher layer information exchanges for assuring their function (e.g., message exchanges between the access control device and remote server). However, there are security mechanisms that control higher level message transfers rather than frames and packets.

An Application-level Gateway (ALG) goes one step further than an IP firewall. Instead of simply checking the IP parameters, it actually looks at application data (e.g., Proxy). It can check elements of protocol to assure that certain patterns are respected or on the contrary are absent). Application gateways add supplementary features to automate the manipulation of firewalls, giving them an access control capability using a narrow set of "Rules." They use RADIUS or TACACS (Terminal Access Controller Access Control System) to communicate with an Authentication Server. An ALG authenticates the user, identifies the device to be accessed, authenticates the access and punches holes in the firewall that are assigned to the discrete IP address of the device and the necessary TCP/UDP port numbers. Firewall rules are automatically created that correspond to the required functions, the technician does his work and then logs out of the access manager. The access manager than removes the rules it just created, thereby removing any holes that could have been left open, hence preventing unauthorized access. The network is then locked down again from the outside until another request comes in for access.

27.3 Designing for Disaster-Resistance

The power delivery system, as other National Critical Infrastructures (CNI), is subject to particular attention and concern relative to their vulnerabilities facing natural catastrophe and intentional damage. They are in most countries subject to state-specified obligations assuring their resistance and their rapid recovery in case of major disasters. The communication system being an essential component for the re-establishment of the power system after any major disruption, it must be particularly robust, geographically redundant, and tolerant to a great number of anomalies. Both the grid automation system and the connection of power system field intervention staff to their operation and support base depend upon the availability of operational communications.

Various threats are to be analyzed through the risk and impact assessment table presented in Fig. 27.1 with or without regional influences (earthquake, storm, flood, fire, intentional and accidental damage, cyber-attacks, system or equipment failures, etc.). Proper mitigation plans are to be devised into the system design, network design or into the processes.

Here, we focus on design aspects of the telecommunication network infrastructure, leaving the specific processes to Part 5.

The telecom infrastructure and service must be designed to:

- Tolerate the loss of infrastructure in a node, in a link or in a region
- Tolerate the loss of mains power supply for a relatively long duration
- Redirect substation communications to a backup Control Center (e.g., destruction or major damage of the Control Centre)
- Include fast deployment communication systems (e.g., radio, satellite) for implementing temporary communication links and networks to replace the damaged or non-operating facilities or to constitute temporary relays for the Utility's restoration staff

Disaster Recovery is not a concept specific to telecommunications but a general plan covering all aspects of the Electrical Power Utility. As such, if telecom assets, infrastructure and staff are located within the perimeter of Utility's premises, then they are integrated into the Utility's Disaster Recovery and Business Continuity Plan (DR/BCP).

However, if telecommunication services are provided by a different entity, with staff and assets at other sites, then the coherence of the telecom Service Provider's DR/BCP with that of the Utility must be assured and periodically audited.

Network Fault Tolerance

It is general practice in operational telecom networks, whichever the employed technologies, to design alternate routing of information in case of a link or a node failure. Automatic service restoration ranges from milliseconds to seconds depending on the service requirements and employed mechanisms. Partitioning of the network into a hierarchical structure and strong mesh structures in the core reduces collateral impacts of failures across the network. Partitioning of large protected switching loops into smaller ones through subnetwork aggregation layers described in Chap. 23 increases the robustness of the telecommunication network against regional failures and in many cases reduces the restoration time in case of failure.

Some common redundancy measures for building high availability telecommunication networks are:

- Duplicated equipment;
- Equipment with redundant modules (Power supply, CPU, service cards, access ports, etc.);
- Physically independent communication media (e.g., optical fiber and radio);
- Different communication technologies (e.g., SDH and PLC);
- Alternative/Mesh network routes;
- Distributed processing systems;

Power Supply Independence

An essential attribute of any operational telecommunication service is its continuity in case of AC power interruption. Although traditional telecom Service Providers are equipped with backup battery facilities, the dimensioning of the latter, associated to the significant rise in traffic solicitation and therefore power consumption in disaster situations, may lead to the unavailability of the vital communications of the Power system. The availability engagements of Telecom Service Providers, as specified in SLAs, are often insufficient to cover this essential requirement.

A dedicated infrastructure is generally composed of telecommunication assets located exclusively (or essentially) inside the electrical sites (e.g., substations), many times equipped with diesel electric generators. Its power supply and backup facilities are dedicated to the utility's equipment (no unexpected load), are dimensioned according to utility requirements, and in general, common with the critical facilities that need to communicate: if there is power for the RTU and Protection relay, then there is power for the associated communication equipment.

Backup Facilities

One of the most important resources in business continuity is having backup facilities, mainly for the operational services continuity. Defining the backup facilities requirements implies, knowing what services are essential, in how much time they shall be restored and for how much time they will be needed. The following requirements shall be considered for backup facilities implementations [Based on NERC Standard EOP-008-1—Loss of Control Center Functionality]:

- Select a safety location specially in terms of natural disasters risk that can be easily accessed;
- Be sure that all needed data will be present at the backup facilities in case of disaster, through storage/database synchronous replication for the most critical information or through data restore for the less critical;

- Include in the backup facilities all the tools and applications that allow visualization capabilities that ensure that operating personnel have situational awareness of the BES, and all those needed for the minimum business/corporate activities;
- Assure all data and voice communications needed for the regular service operation, including voice and control communications to the critical substations and power plants, and to communicate outside the organization including Internet access;
- Include reliable power sources such access to redundant distribution power lines, diesel generators, diesel refill contracts, etc.;
- Be sure that all the physical and cyber-security requirements applied to the main facilities and control centers are also guaranteed;
- Implement an operating process for keeping the backup functionality consistent with the primary control center.
- Do not forget to assure food and medical first aid;

As all of the components of the BCP, the backup facility availability and functionality must be tested in a regular basis. However, if cost effective, its usage as a "hot site" is recommended, for example in a distributed processing architecture or under hot-standby architectures.

Fast Deployment Communication Systems
Provisioning fast deployment communication systems such as radio and satellite, allow implementing temporary communication links and networks to replace the damaged or non-operating facilities. It also is an efficient manner to constitute temporary relays for the Utility's restoration staff coordination in the damaged area. Satellite data capacity can be procured for a certain annual duration allowing the transport a mobile communication cabin to the disaster site and establish communications to the remote coordination and support base.

Part V
Maintaining Network Operation

Maintaining Network Operation—Introduction 28

Operational information exchange in the electrical power utility is undergoing tremendous growth, leading to an ever more complex and extensive communication network as described in the previous parts of this book. The corresponding operation and maintenance necessitate specific processes and tools well beyond the ad hoc management previously employed in most cases.

The new operational communication network is also employing a much wider range of technologies which moreover evolve much faster than before hence requiring a great diversity of skills: the field worker can no longer be considered as autonomous and increasingly needs to be supported remotely by a wide range of technology experts. These experts can be located in one central facility or increasingly dispersed across the wide footprint of the communication network. They can be in-house or part of an external contractor or vendor organization. Processes and tools need to be adjusted accordingly.

The adopted processes, tools and field communication systems depend also upon the organizational aspects of the power company: fusion or separation of operational and corporate enterprise services, of IT and telecom systems, and of local and wide area communications, as well as the adoption of in-house or contracted operation and maintenance services with specific security issues in the latter case.

This last part of the book is dedicated to the analysis of the new operating practices, processes and tools for maintaining this transformed operational telecom network. Service management is largely discussed under Part 3 and deploying the network and associated tools in Part 4. Here the focus is on maintaining the network itself although reference is frequently given to previous discussions. The material is largely based on the work performed by CIGRE WGD2.33 providing a comparative study of experiences and targets among utilities for the day-to-day operation and maintenance of their telecom infrastructure.

© Springer International Publishing Switzerland 2017

C. Samitier (ed.), *Utility Communication Networks and Services*,
CIGRE Green Books, DOI 10.1007/978-3-319-40283-3_28

The scope is limited to the communication infrastructure covering utilities' sites and operational assets excluding therefore customer premises (e.g., metering device and metering infrastructure) whose management and maintenance is in many respects very different, mainly due to the number of concerned sites being orders of magnitude larger than utility operational assets.

Our scope covers operational network management both in normal time and in disaster situations. Disaster preparedness was covered to some extent in the previous part as a network deployment issue and shall be briefly discussed here as an operational issue. Although there is no standard border between normal time and disaster-time O&M and no specific criterion for switching over to a disaster-time process, here we define the boundary between the two as follows:

- Normal O&M is the process of maintaining the communication service for **optimal cost and performance** using a **time-sustainable** set of processes, workforce and tools. This includes dealing with normal incidents of the communication network with the target of **optimized operation**.
- Disaster O&M and the associated recovery, on the other hand, consists in managing major incidents with "disastrous" impact on the service in order to maintain **service continuity** for the most critical services using **special processes**, and through mobilization of **extra people and extra tools** (e.g., from other regions) as compared to those employed for normal operation. The target is no longer optimality but rather the **survival of essential services**.

Reasons for a Formal Approach to O&M 29

Any attempt for providing a formal process governing the day-to-day tasks and their corresponding interactions in any human organization, inevitably creates apprehension. The employees whose action is formally modeled or invited to change behavior according to a formal process often perceive this as extra, unnecessary tasks adding to the already heavy workload. Moreover, they apprehend the process change and its monitoring as an intrusion and a means of policing their actions with consequent unfair appreciation of the quality of the performed job. The employer, on the other hand, based often on the outcome of previous experiences, may consider process modeling as unproductive investment, with no perceivable improvement at the end of the day, and finally more staff needed for the same effective workload.

With such a severe account on work process modeling, there are still many good reasons for undertaking the analysis of the operation and maintenance process for telecom networks and services in the electrical power utility. It may therefore be appropriate to start our analysis with its justification presented in Fig. 29.1.

Hereafter, some major points of Fig. 29.1 are described:

Assessment of Service Provisioning

The power utility in the new liberalized market is constantly in the quest of cost reduction. The reliability of operational communication services and their cost of achievement are increasing challenged by alternative service provisioning modes. It is essential to have a trans-Utility reference O&M processes and associated Key Performance Indicators (KPI) for meeting operational service imperatives.

- for fair assessment of alternatives (in-house/external)
- for determining market's capability to deliver appropriate O&M services
- for consequent partitioning and interfacing of in-house and external service provision

© Springer International Publishing Switzerland 2017

C. Samitier (ed.), *Utility Communication Networks and Services*,
CIGRE Green Books, DOI 10.1007/978-3-319-40283-3_29

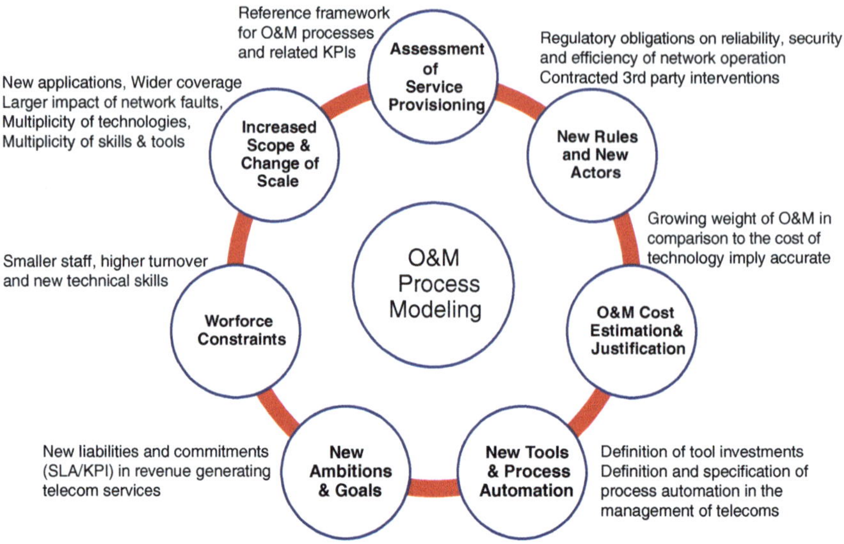

Fig. 29.1 Reasons for modeling the O&M process

Increased Scope and Change of Scale

New power system applications are bringing a great number of new communication service requirements in the substation and across the grid. These applications range from the control and maintenance of the distribution grid, the integration of renewables and outage prevention to HV grid stability control and new system protection schemes. This explosion of communication-based intelligent applications is changing the scope and scale of O&M activities. It is also accompanied by the introduction of new communication technologies and the evolution of the telecom networks necessitating multiple skills and tools to maintain the system.

New Rules

Another major on-going change concerns external regulating authorities and legislation on efficiency of network operation. The power utility is expected to keep its infrastructure in good operating condition and its operational expenditure including those related to O&M need to be precisely determined and justified in order to be able to reflect it in the price of delivered electrical power. Moreover, the increased weight of electrical power utility as a Critical National Infrastructure due to increasing dependence upon power supply increasingly results in sanctions and penalties for delivery failure. This, in turn, increases the importance of communication reliability and consequently the promptness of the O&M.

New Actors and Players

From an incumbent model of vertical power utility where all components of the power production and delivery were part of a same company (and therefore all related communications were internal), the power industry is moving to a partitioned scheme leading to the multiplication of partners requiring exchange of information and consequent multiplicity of technical issues and solutions. The O&M process must now cover such issues as cyber-security and protocol compatibility for external communications. Moreover, the creation of new energy corporations covering multiple utilities in multiple countries is leading to uniform O&M practices, tools, and processes.

The offer for external provision of communication services to utilities or for the operational management and maintenance of utilities' dedicated telecom infrastructure has become abundant. However, the gap between utility operational needs and mainstream telecom operator products and services, has also become larger than ever. The reduced dependability of public or enterprise-type telecom services (e.g., short autonomy in case of power interruption) has opened up a space for the deployment of dedicated "blue-light" communication services (e.g., mobile services). Assessing the suitability and cost-effectiveness of these specific offers requires models and references not only of the final quality of service but also the way to ensure these qualities (e.g., O&M processes).

New Workforce constraints

In-house capability of electrical power utilities is shrinking due to aging workforce, downsizing policies and economic constraints of organization, multiplicity of technologies and technical issues (old technologies persist while new ones are introduced into the network), and the difficulty to attract and maintain appropriate skills. The multiplication of technical issues and solutions, the fast rate of change of technologies, and the fast turnover of technical teams both in-house and outsourced, is leading to a reduced grasp of the network and its potential problems. It is increasingly important to have a formal and specified manner to operate and maintain the network.

New Tools and Process Automation

Mainstream telecom and IT world is proposing a wide range of tools to the utility telecom networks for automating the processes of network and service management. These tools are designed and optimized for public service providers and enterprise networks. Rather than modifying O&M processes according to market tool capabilities, an assessment of O&M practice in operational context and its required evolution allows the implementation of more suitable and cost-effective tools according to operational requirements.

O&M Cost estimation

The evolution of telecom technology is increasing the weight of O&M in the full life-cycle cost of the system. Analysis and modeling of the O&M process allows cost optimization and justification (compared to the consequences of not maintaining).

New Ambitions and Goals

Utilities aiming external revenue telecom activity need to coordinate SLA/KPI commitments with their operational objectives and determine formal priority arbitration as part of their process. This necessitates a clear definition of the operational O&M process.

O&M Scope, Process, and Organization 30

Implementing a telecom operation and maintenance organization consists in defining and partitioning of tasks, allocating these tasks to roles in an organization, defining processes, and tools for the fulfillment of tasks and their continuous monitoring, in order to provide telecom services of optimal quality and cost to the end user.

There is no unique strategy and no standard dimensioning solution for the O&M functions. Some questions that need to be considered are:

- What is the size and scope of the telecom services to deliver, and for how many users?
- How wide and how complex is the telecom network?
- How high is the expectation on the quality of service?
- How low is the limit of "reasonable" cost?
- How many people can constitute the workforce and with which capabilities?

A small and relatively informal process perfectly adequate for delivering telecom services to a small group of operational users (e.g., SCADA and Protection) cannot be applied to a telecom network serving the entire power corporation. A large and complex process designed for telecom operators, on the other hand, cannot be applied upon a small operational telecom O&M team in the power utility.

Operator-grade management tools have cost and effort requirements not justified for most utility operational telecom networks. On the other hand, the service continuity and quality monitoring requirements of the operational user are increasingly beyond the capability of the "no-tools" situation of many power utility operational telecom networks.

The success of a management set-up depends upon the adequate dimensioning of the organization, process, and tools relative to the scope of activity. Determining the management scope and perimeter and analyzing its evolution is therefore a key issue before undertaking the assessment of tools, processes, and organizations.

© Springer International Publishing Switzerland 2017

C. Samitier (ed.), *Utility Communication Networks and Services*,
CIGRE Green Books, DOI 10.1007/978-3-319-40283-3_30

Among the determining factors for a suitable O&M the most important are the scope and perimeter of the telecom network and the communication service, the nature of service provider to service user relationship, and the size and scale of the operational organization. A redesign of the management system and related processes is often justified and triggered by the intention to change any of these factors beyond its initial deployment objectives:

- A sharp increase in the size and scope of the telecom network
- Significant increase in the number of services or of service users
- Major change in Utility organization (e.g., separation of telecom entity)
- Major change in service commitments, in liability, or third-party involvement
- Sharp change in technical staff capability (new technologies, retirements, …)
- Deployment of process automation tools and change of roles and tasks

It is worth noting that very often a review of the O&M is the forgotten component of major network transformation projects.

30.1 User-Provider Relationship

In part 2, Chap. 11 we discussed different service provisioning models leading to different user-supplier relationships as presented in Fig. 11.1. We will employ these same models in our discussion of operation and maintenance.

In an embedded delivery scheme (Mode A in Fig. 11.1), which is still the most commonly encountered, the network is composed of electrical substations and control centers although some other adjacent administrative sites such as engineering and project offices may also be covered without any determining role in the network footprint or its operation. The operation process is trivial and implicit to the small scale, simple and time-invariant nature of requirements, informal user relationships, and non-recovery (sometimes non-assessment) of costs. Such an embedded telecom network in the power system's operational organization is often accompanied with separate corporate enterprise communication services generally delivered by an external (public) service provider or outsourced to an external contractor under the responsibility of the IT department with separate organization, O&M process and tools. Service delivery being limited to a small number of users, the O&M tasks are often performed by a reduced team of in-house staff with mainly technical skills and a rather informal service management process.

Fusion of operational and corporate communications in a same telecom network and therefore operated through a same set of processes and by the same team is often a consequence of one or several of the following factors:

- Extensive capacity and capability in the operational network infrastructure
- Growing interactions between the utility enterprise and Utility operational users
- Growth of enterprise communications and consequent cost of external service provisioning

- Cost saving through a single network infrastructure and O&M organization for both corporate enterprise and operational services
- Deployment of service convergence technologies (e.g., IP) and borderline applications (e.g., field worker communications, asset monitoring)

Delivering both operational and corporate enterprise communications from the same network (and by the same team) generally results in the extraction of the telecom activity from the power system operational organization and moving it from mode a to mode b, c, or d (refer to Fig. 11.1) depending upon the size and scope of the company and of the required services. The network may be run through an in-house O&M organization and process or by a contractor delivering services through utility-owned telecom assets but through its own O&M process. The more we move from an in-house delivery to the usage of external contractors, the more we need formal interactions and processes, replacing technical capability by contract management and formal process skills.

30.2 Network Perimeter for O&M

Different domains constituting the potential perimeter of telecom O&M in a utility organization is shown in Fig. 30.1. The extent of coverage of this perimeter by a single O&M organization is variable among power utilities. It may include or exclude substation LAN infrastructures, control room IT LAN infrastructure, teleprotection signaling and interface/protocol converters. The resulting O&M, necessary processes, skills, workforce, and tools can vary substantially.

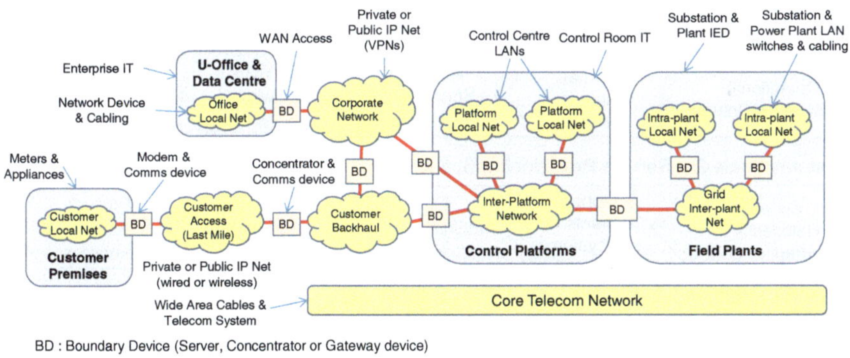

Fig. 30.1 Network domains in the electrical power utility defining telecom O&M perimeter

30.3 Scope of O&M Activities

The scope of telecom O&M depends indeed on the perimeter issues developed here above, but also on the service users (only operational, all internal users, or internal and external users), external contractors and suppliers to be managed, cost recovery scheme and liability structure. The service delivery process described in Part 3, Chap. 17 provided an overview of the different "operation" activities of the telecom service delivering entity. In particular, a matrix of current operations was presented in Fig. 17.2 covering all tasks that need to be fulfilled in the largest, "most mature" scheme of telecom service delivery. However, not all these tasks exist in every power utility. Figure 30.2 hereafter presents a business maturity model for telecom service in power utilities. The scope considerations in this figure will largely impact the size and scale of the O&M activity.

Maturity Scale 1 (Service Delivery Scope) determines the perimeter of service users. Progressing on this scale reduces user-provider proximity and the uniformity of service objectives. At the left extremity we have a dedicated network operated for, and according to a single user (power system operation) requirements. Moving rightwards, the telecom network operator increasingly needs to consider multiple network users with potentially conflicting interests. Prompt operational service impact analysis on network faults and continuous monitoring of SLAs allow appropriate management of conflicting priorities.

Maturity Scale 2 (Service Provisioning Scope) determines the extent to which the telecom service provider depends upon his service suppliers and contractors. Progressing across this scale increases the requirement for formal agreements and

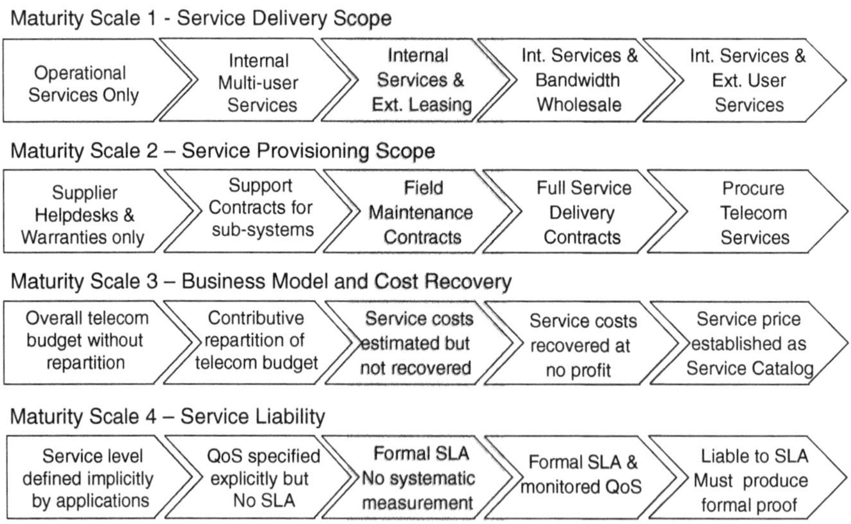

Fig. 30.2 Business Maturity scale model applied to the scope of O&M activities

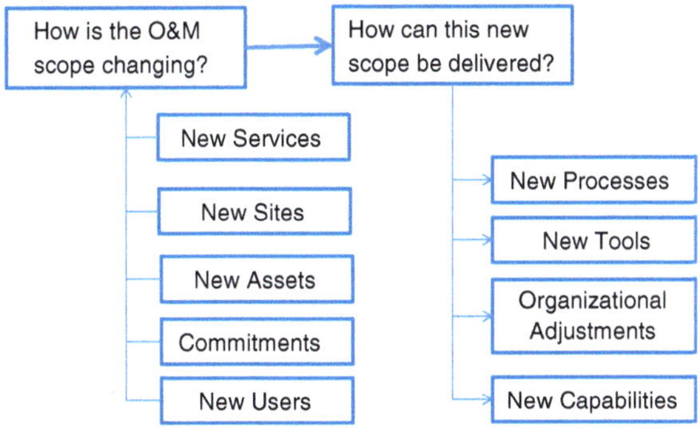

Fig. 30.3 Change of scope in telecom O&M

measurable objectives as well as tools and processes to monitor and manage the relationship.

Maturity Scale 3 (Business Model and Cost Recovery) determines the necessity to measure the usage of network resources by each service and consequent metering and account management. Nonmeasured usage of network bandwidth and non-recovery of service costs restrains proper estimation of user communication requirements.

Maturity Scale 4 (Service Liability) finally determines the extent to which the telecom service provider is committed to monitor the quality of service delivered to different users.

It should be noted that the objective of power utilities is certainly not to move as far as possible across the maturity scales. The positioning across the scales is a question of requirement, company policy, network size, cost considerations, etc. However, it is equally important to note that the positioning is not invariant but moving in time. New business situations necessitate adjustments in the way the network is operated and maintained as presented in Fig. 30.3 hereafter.

30.4 Evolution of O&M Scopes and Processes

O&M process and organization follow the changes of scope, perimeter and scale of the power utility telecom activity. From an initial state of informal and implicit management of a low capacity infrastructure dedicated to time-invariant applications, at the bottom of the "maturity scales" shown in Fig. 30.2, the management requirements evolve together with the user base, service requirements, and network technologies. Any transformation plan changing the scope and size of the network beyond its initial deployment objectives must be accompanied by a review of the management requirements and a redesign of O&M process and organization. The deployment of new tools, although often part of the process redesigns, can in no way provide on its own a full solution for running the transformed network. Significant change-triggering conditions are listed in Fig. 30.4 hereafter.

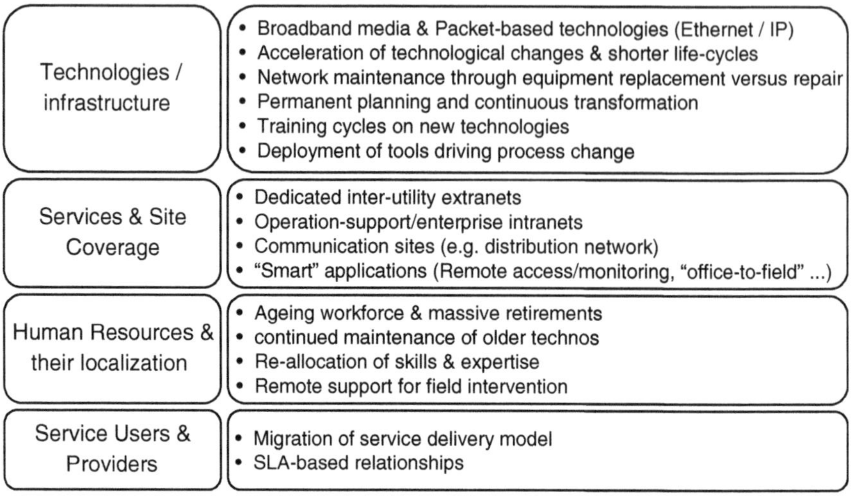

Fig. 30.4 Change triggering conditions for power utility operational telecom O&M

30.5 Transforming the O&M

Power company telecom organizations are at different situations regarding the described change factors and the resulting O&M maturity. In the same manner that Fig. 30.2 provides a Business maturity scale on the telecom service delivery in the power utility, the evolution of O&M capabilities and process complexities can be modeled into a maturity scale composed of 5 stages situating the organization in terms of present and future management practice:

1. Informal—corresponds to ad hoc operation with a small and co-located "multi-purpose" team operating telecom services by case-by-case response to problems as they occur and are detected.
2. Repeatable process—Case-by-case response of the previous stage is replaced by regular patterns and roles established in the day-to-day operation of the network. Service management remains however in the reactive mode (reacting to monitored network alarms) by coordinated teams.
3. Formally defined—All operations of the O&M team are formally documented processes known by all actors allowing more proactive (predictive and preventive) actions on continuously monitored network performance and availability. A standard formal framework such as ITIL may be used for specifying roles and interactions.
4. Managed operation—O&M process effectiveness is monitored and measured through adequate KPIs (Key Performance Indicators), such as customer satisfaction and incident duration statistics and trends.

5. Optimized and Automated—Formal processes based on best practices are automated for optimizing the efficiency and the cost of the O&M operations.

In the same manner as for the business maturity scale, the O&M process maturity model described above may lead to the wrong impression that the objective of the utility O&M organization would be to mount continually the steps of this scale producing ever more formal and automated processes. In reality, depending upon their specific contexts, power utilities do not necessarily intend or require evolving to a high degree of telecom O&M complexity. It is more important to assure that the adopted processes are effectively being followed with conviction of their benefit rather than progressing along the scale but constantly working around the process because of the time and effort requirements beyond what the organization can provide. Process effectiveness and the extent of its application must be periodically evaluated to allow continuous improvement through adjustments of the actual operation and of the process itself.

Moreover, a process involves people, organizations, relationships, and interactions which cannot always be changed overnight. Changes can produce fears, reactions, and resistance unless understood, accepted, and supported by the concerned staff who must preferably be involved in their elaboration, together with an adequately budgeted migration plan including training, simulations, and tools.

At present most power utility operational telecom organizations are at step 2 with roles and responsibilities precisely defined and distinguished, although in some smaller teams O&M roles may be adapted to day-to-day requirements (maturity step 1).

In-house staff's time and effort in many utilities are not measured and there is often no real cost accountancy and repartition for internal workforce. Telecom O&M workforce may consequently be solicited for assistance to operational user departments on service impact issues, performing tasks sometimes beyond their perimeter. A fair level of formal definition (maturity stage 3) currently applied to in-house workforce in some utilities, is a documented description of daily, weekly, monthly, and annual actions to be performed. However, quantitative monitoring and measurement of O&M performance through key performance indicators (stage 4) is mainly applied to external suppliers and service contractors with whom the interactions are contractual and formal. Utilities with more external contracting are consequently the ones often with the highest level of formal process.

Automation of processes concerns all machine-aided automatic interactions across the operation and maintenance organization. It permits automatic treatment of user requests, generation of work orders, notification of contractors, etc. for increased efficiency (stage 5) and is widely deployed in IT and service provider contexts constituting the main driver for formal frameworks. It may comprise user/provider relation aspects, such as automatic generation of alert messages, notifications, and service desks, but also resource management processes, such as incident ticketing, assignment and escalation, report generation, inventory, and configuration applications as well as statistics and trend analysis. Companies with merged IT and telecom O&M are naturally the ones who have gone the furthest in process automation using essentially the tools and applications which are thought and designed for the IT world. Automation is much less developed in smaller organizations missioned to operate and maintain an operational telecom network due to the rarity of smaller adapted off-the-shelf tools on the market: potential benefits in terms of reduced effort and time do not necessarily justify developing custom applications considering the scale of the service delivery. A number of smaller platforms targeting this segment are however appearing in the market.

To sum up, O&M processes, their extent of formalism and the way there are implemented, manually treated or automated, must evolve in order to assist the telecom O&M organization in performing its missions more efficiently.

However, for any scope and size of mission, there is a maximum point in formalism, role partitioning, and automation, up to which the organization can increase its effectiveness, but beyond which, pushing further renders the actors less efficient and increases the risks of error and loss of control.

The process for dealing with day-to-day tasks is not the same as the one for dealing with emergency situations for which more simplified processes should be foreseen. The process for dealing with a stable network cannot be the same as the one in force during important transformation works. Formalism is about acting in a predetermined repeatable manner but cannot cover every situation whatever be the thickness of the process. The organization must also rely upon human intelligence navigating between formal and informal, automated and manual according to contexts and situations.

30.6 Operation and Maintenance Organization

The size and scale of the O&M organization and the nature of tasks that need to be performed differ substantially depending on the perimeter of the network, number of users and services, and the scope of work. Sites are generally dispersed over a large area rendering the access to sites a dimensioning factor. Field works constitutes the largest part of the O&M staff and the part which is most likely to be externally contracted or shared with other power system field works. The more the network is extended geographically, the more the field workforce is likely to be shared (with other power asset maintenance) and/or contracted.

Figure 30.5 illustrates in general terms the functions that will need to exist in each of these situations. The generic scheme defines a network management organization composed of network supervisors, maintenance and support, service managers, and network planners (Transformation Engineers). Each group can indeed be subdivided according to the size of the organization or may be inexistent depending on the scope of work of the organization. Obviously, the more these splits are formally defined, the more the process needs to formalize exchanges between actors causing further overhead and reduced flexibility but at the same time more precise definition of responsibilities and focused dedicated resources.

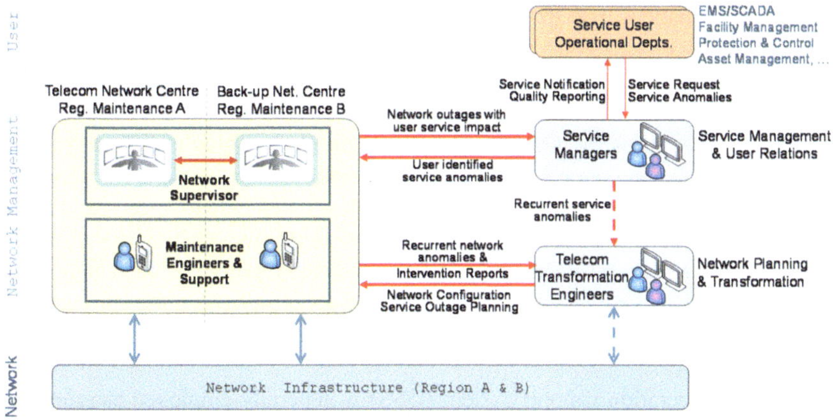

Fig. 30.5 A generic management organization and process for telecom O&M

For example, where communication services are procured from an external telecom provider, the tasks of supervision, maintenance, planning, and transformations of the telecom infrastructure are indeed inexistent in the utility organization. However, the external provider shall not directly deal with the final service users across the power utility but rather with a telecom service manager who represents the power utility as a single customer, looking into the quality and cost of the delivered service, and checking the respect of contractual commitments, and service level agreements (SLA). The telecom service manager shall act as the interface between utility communication service users and external telecom service providers, compiling requests, and detected anomalies to inform the provider and negotiating service extensions and upgrades. The tasks to be performed are much more those of contract management than technical tasks.

At the other extremity, a power utility that applies no external services and commodities for the provision of telecom services shall have an important telecom network infrastructure to run. It shall therefore have one or more network supervisors looking after the proper operation of the telecom network, maintenance engineers, and technical support in order to maintain the network, planning, and transformation engineers to look after the extensions and upgrades, and a service management activity to assure user satisfaction and interfacing. The service management task is more on the technical side and requires less contract management and legal effort than the former case.

The great majority of practical situations are in an intermediate situation tending more toward one or the other of the extremities. The telecom organization has some infrastructure to run, and some external service providers/contractors to manage in order to deliver communication services to the utility users.

30.7 Network Operation Center Activities

Telecom O&M tasks are generally conducted from an independent dedicated Network Operation Center (NOC) most often located in a utility power system control facility unless the O&M is fully contracted, in which case it may be performed from the contractor's premises. The operations may be hierarchical, through regional and national level centers or centralized depending upon the size and dispersion of the infrastructure and service users. Moreover, multiple functional levels for O&M do not necessarily signify dedicated people in each region. The O&M can act as a single team in one location for centralized supervision and have access to regional intervention teams either dedicated or shared with other intervention activities.

The term operation support system (OSS) is generally employed to refer to the online and offline information systems and platforms used for supporting the technical operation and maintenance of the telecom network, as well as the activities, processes, and interactions for current problem solving, incident management, and service monitoring. The current operations process chart presented in Fig. 17.2 (Part 3) is derived from the process framework of NGOSS (Next Generation OSS) which also includes an information framework (or model) and an application framework.

In the following sections, we shall discuss not only the tools and information systems but the specific activities and processes in the telecom O&M of the electrical power utility.

Service management and user relation activities having already been discussed in Part 3, Chap. 18, in this part we focus on the technical network infrastructure, covering fault and incident management at the NOC, configuration management of the network and of its equipment, as well as the monitoring of the network infrastructure performance.

Activity	Recurrence frequency
Creation of new circuits	From weekly up to daily
Receive and Acknowledge alarms	From daily up to hourly
Open incident and Assign Intervention	From weekly up to daily
Performance check	From less than monthly (e.g. during preventive maintenance) to daily scheduled and up to continuous (polled every few minutes)
Remote configuration	From weekly or daily (up to hourly)
Preventive maintenance for critical or older equipment, ventilation filters, and power supply batteries)	From every 6 months to over 2 years
Server redundancy checks, log purge, data consistency, etc.	From monthly to weekly

Fig. 30.6 Frequency of recurrent day-to-day operations in present day utility telecom

Some recurrent actions of the telecom O&M and their typical recurrence frequencies are given in Fig. 30.6 hereafter. The recurrence frequencies are an indication of the volume of activity related to each type of task, and therefore, the relevance of automation and tools in each case.

31 Managing Faults and Anomalies

Detecting and locating network anomalies and performing appropriate corrective action are probably the most fundamental activities of the O&M organization. As it can be seen in the table of task recurrence in Fig. 30.6, it corresponds to the most frequent daily actions to be performed. The recurrence becomes even higher during major disruptions, such as natural disasters and network transformation projects.

Network anomalies include equipment or line failures as well as subnormal operational conditions or performance (data errors, transmission time-out, received signal level degradations, traffic saturations, degraded environmental conditions, power supply, etc.). Anomalies are detected by communication devices or specific monitoring devices through continuous observation and field detection of values beyond preset thresholds. Fault management is a subset of a wider domain called event management covering both equipment and network faults as well as other performance-related events.

As illustrated in Fig. 31.1 any fault management system is composed of three constituents: the field device (either the communication equipment itself or a remote management unit), the data communication network for exchanging management information, and the processing platform for fault and event data.

31.1 Fault Detection

Fault detection through Power System SCADA

The simple historical way to detect telecom faults has been the usage of the power system SCADA RTU to bring the alarm contacts of telecom equipment to the control center. This approach is based on historical hypothesis that the telecom network is part of the substation facilities and under control of the power dispatch center which will inform telecom technicians of any fault. Not only this hypothesis is no longer true, but also it blocks the organizational evolution and the scalability of the network through its technical/architectural limitation. Moreover, this

© Springer International Publishing Switzerland 2017

C. Samitier (ed.), *Utility Communication Networks and Services*,
CIGRE Green Books, DOI 10.1007/978-3-319-40283-3_31

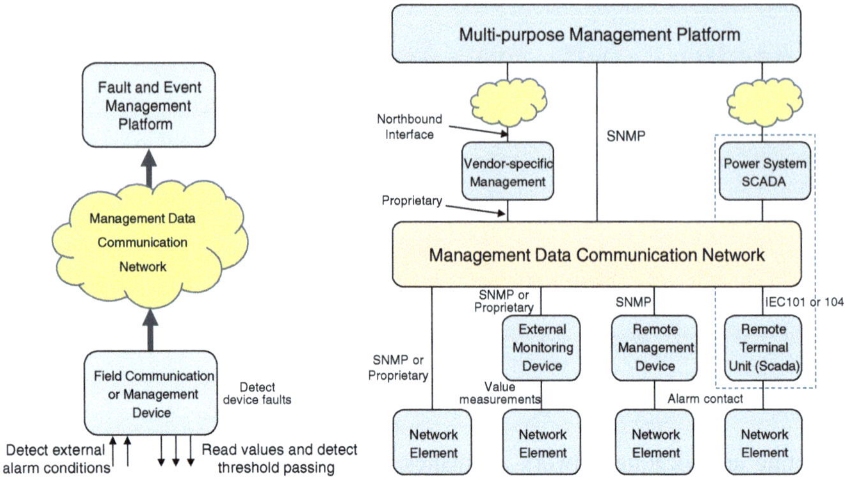

Fig. 31.1 Constituents of a fault management system

approach burdens the data volume of the SCADA system, with its more stringent real-time constraints, with supervision data having no real-time requirements.

Fault detection through external condition monitoring

Telecom subsystems with no self-monitoring intelligence often use an external condition monitoring system in the same way as other primary power system assets such as transformers and circuit breakers. The most typical example is optical fiber cables monitored by optical time-domain reflectometers (OTDR) dispersed across the network at selected cable nodes continuously monitoring the attenuation and discontinuities of the fiber cables (see Fig. 31.2). When a change is encountered, an embedded communication device transmits a message to the central platform. Similar operation allows monitoring equipment room temperature, power supply output, etc., detecting anomalies and notifying the center. SNMP (Simple Network Management Protocol) discussed in the following sections is a very convenient manner for performing these simple monitoring operations.

Fault detection through vendor-specific NMS

The most common way to detect telecom network faults is the vendor-dedicated proprietary network management systems (NMS) associated to particular type of equipment, giving not only indication of alarms but also the capability of in-depth diagnostics and investigations for the equipment or for a single-vendor subnetwork. However, the multiplication of types of equipment in a utility telecom network often leads to several platforms (typically 10-20 in a large network). This includes both functionally equivalent equipment from several vendors and different functional blocks (e.g., transmission and multiplexing equipment, data switching and routing, voice platforms, etc.). A same event in the network shall generate alarms

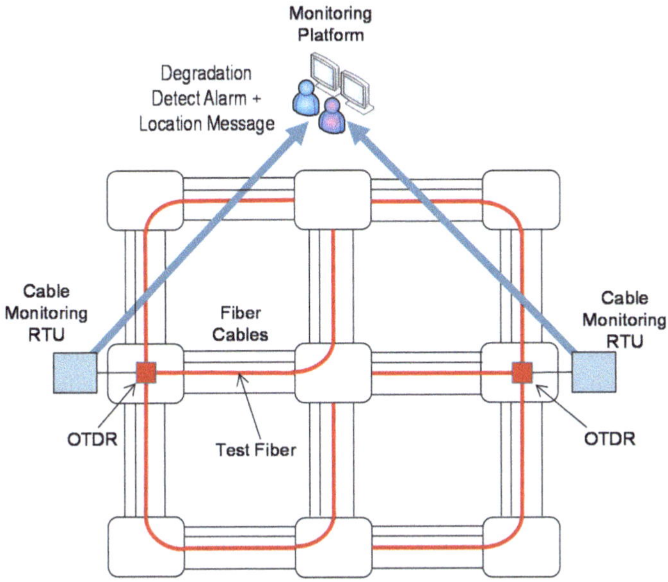

Fig. 31.2 Cable fault detection and localization through external condition monitoring

and notifications of different format and different content in many distinct management platforms.

Fault detection through a unified platform
Multipurpose standard SNMP-based platforms allow multi-vendor and multi-technology fault management providing a unified view of all network events. Each SNMP device includes a structured management information base (MIB) containing all monitoring and management data relating to the device which can be extracted (GET) or modified (SET) from the management platform. The device can moreover emit state changes (TRAP) to indicate an event or alarm to the platform. The great majority of the new generation network components are SNMP-enabled either natively or through an SNMP agent integrated into them. For older equipment there is always the possibility to install an SNMP remote management unit converting alarm contacts or serial port data into SNMP for transmission to a central platform.

Vendor-specific or Unified platform?
Power utilities mostly use vendor NMS of different subsystems (SDH, access, etc.), market standard SNMP collectors, and enterprise network management platforms for IP networks sometimes used also for monitoring MIB elements of other devices such as manageable rectifiers.

It should be noted that there is no opposition or contradiction between the two approaches. A vendor-specific NMS platform allows a more in-depth analysis for a particular type of equipment management while a unified SNMP platform on the other hand, enables fault correlation and root cause analysis, but not the same capabilities of in-depth investigation and detailed diagnostics. Vendor-specific management tools are being employed in a many utilities for analyzing faults, while many have also deployed SNMP platforms for detection and localization of faults across multiple subsystems. The two approaches are complementary and can be associated into a same fault detection and localization system.

Hypervisor or "Manager of managers" approach

A unified management platform can also receive management data from network elements through vendor-specific management platforms. The unified platform in this case will be used only to visualize the overall network and to perform higher level tasks such as trend analysis and statistics. The interface of the vendor-specific management system to a higher level platform is called the Northbound Interface (NBI). Currently, SNMP and Web service (HTTP) are the most common ways to implement northbound interfaces. Using Northbound Interfaces is an adequate approach when network elements only support vendor-specific management. Often a very limited amount of the available information in the vendor-specific management system is required for the unified manager for detecting faults and anomalies corresponding to major and minor alarms.

Managing network faults exclusively through individual management systems indicating faults to the overall manager creates a common mode failure point—that is to say, a power failure in a network management system, for example, will block all management information to reach the overall management platform.

Detecting dead, inaccessible, or malfunctioning device

The principle of detecting a device fault through its own auto-diagnostics or notification presents the problem that a dead or not powered device will not notify its state, and a malfunctioning device may not be aware of its own anomalies or may not have access to management communications to notify its detected problems.

Some telecom devices (e.g., Ethernet switch) may have a "dying gasp" feature which is the capability to send a last signal in the event of power outage storing enough power for a brief period for sending the message without external supply.

Another feature existing in digital transmission equipment is the alarm indication signal (AIS) allowing other devices around the anomalous device detecting the problem to transfer the fault information (essentially for communication recovery and resilience) and consequently to signal it to a fault management platform. However, this solution also results in multiple fault indications to the management platform for the same fault requiring some sort of alarm de-correlation to be performed in the fault management platform.

A more general solution to the detection of dead or noncommunicating nodes is the usage of periodic polling of managed devices in addition to the event-driven alarm (or SNMP TRAP) mechanisms. A periodic IP Ping of all addressable devices in the network allows the management platform to check the communication capability and accessibility of network elements. In this case, it is essential to distinguish between equipment in major alarm status and non-responding network elements.

31.2 Fault Localization and Problem Management

As described here-above a fault occurring in the telecom infrastructure can cause multiple alarms in different network elements in the same logical layer (e.g., assets facing the faulty element) and in other logical layers (e.g., assets relying upon the faulty element). A transmission equipment fault can, for example, cause transmission layer alarms, multiplexing layer alarms, switching element alarms, etc. A "problem" is an unknown "underlying cause" of one or more signaled anomalies (i.e., incidents). Alarms may moreover be repeated many times in a short period of time. Prompt filtering, de-multiplication, and localization of the principal cause of this avalanche of alarms known as Root Cause Analysis (RCA) requires determining any correlation between them and is essential for rapid recovery and for assigning only the relevant technicians or teams to the resolution of the problem.

At present, in the great majority of cases no specific tools are being used for assisting root cause analysis. Although faults are being detected very rapidly using currently deployed telecom fault supervision systems, the localization process is time consuming—1 to 4 h reported by some utilities and a dimensioning constraint in fault recovery cycle by others. Some companies have dedicated "network analysts" performing RCA manually and recurrent faults are being identified and documented for simplifying diagnosis in some utilities. Manual RCA should be understood as analyzing the reported alarms over the disjoint documentation of the network and logical configurations.

Simple overlay network models (e.g., multiplexers and switches communicate over transmission equipment which connects over a fiber connectivity composed of cable segments, etc.) and relatively straightforward dependence relationships fed into an event management system can greatly automate the task of fault localization and assist the network operator in visualizing the fault and in performing fault localization without detailed knowledge of the network.

31.3 Fault Notification and Reporting

Certain critical faults detected by the network supervision may need to be notified automatically to the on-duty maintenance engineer either for immediate action or due to "working hours" operation of the supervision center. Otherwise, faults are reported by the network supervisor following the identification and localization process as part of the incident management.

The proportion of network faults detected through telecom supervision and through impacted users is variable (100-0 to 80-20). This proportion gives a measure of proactive or reactive behavior of the O&M organization: detecting network faults before being perceived in the quality of communication service delivered to the end user. The network supervisor must have a means of integrating anomalies reported by the user into the incident recovery cycle discussed under incident management hereafter.

31.4 Fault Diagnostics

Detailed diagnostics are best performed through vendor-specific NMS of different subsystems, local craft terminals, and some other specific tools, and instruments. Diagnostics can be performed locally at site, or remotely from the network operation center. Remote diagnostics is increasingly required due to time saving and due to no staff at or around the field sites raising the problem of secure remote access to field assets. A useful precaution in this context is to centralize the access of all vendor-specific tools into a single operator interface concentrating hence the security measures and validations at the access point into a single framework.

31.5 Fault Recovery and Reporting

Fault recovery is the process initiated by the network supervisor through assigning a team or a person to the permanent resolution of a problem (or at least finding a workaround if major transformation action is found to be necessary). It can be performed remotely (e.g., from the network operation center) or at site. The recovery time depends indeed on the nature of the intervention, on the travel time, and logistics such as the access to spare parts (2–4 h to a day depending on number and location of stores and possible contracts with suppliers). Site access and travel time are indeed the other critical aspects being often the cause of outsourcing or sharing across the utility of site interventions. Reducing the required skills at site and remote assistance of field workers are some directions taken for mitigating these issues.

32 Incident Management and Work Assignment

Incident Management is the process of handling faults and anomalies detected through the fault detection process or signaled by users (as already described under fault management), assigning adequate resources to their resolution, coordinating the interactions of different actors and keeping track of the performed actions, and finally generating statistics on their frequency of occurrence, resolution times, etc., in order to improve the network or the process (Fig. 32.1).

When communication services are delivered through internal resources (i.e., dedicated infrastructure and in-house O&M workforce) then service and network incidents can be treated in a merged manner.

Even with external network resources and O&M workforce, the power utility still needs in-house Service Incident Management. In this case, a service trouble ticket process, assigns the service contractor for resolving the problem. The service contractor will then handle the network incident through another incident management process.

Service Incident Management comprises essentially the creation and closure of Trouble Tickets, a work assignment and escalation mechanism, resolution time estimation and manual or automated statistics building and dashboard generation.

Some utilities limit the detection of incidents to business hours (no 24/7 attendance) and use network redundancy to cover their service continuity. Outside business hours, the control center can handle the most critical incidents necessitating either the reception of all critical alarms all the time or *an automatic fault notification message conditional upon time-of-day.*

© Springer International Publishing Switzerland 2017

C. Samitier (ed.), *Utility Communication Networks and Services*,
CIGRE Green Books, DOI 10.1007/978-3-319-40283-3_32

Trouble Ticketing	◦ Create trouble tickets ◦ Assign incident resolution to a particular person/department ◦ Keep track of the incident with possible escalation ◦ Close incident and perform statistics on resolution times ◦ Formalized incident reporting ◦ Monitor & report on incident resolution
Work Order Management	◦ Provide support for planning and adjustments of work schedules by automatically extracting other work on complementary pair, including work on redundant routes. ◦ Determine the necessity of checks on service soundness including absence of failure on redundant routes prior to the commencement of planned work. ◦ Support for control offices to regulate work and/or manage the progress of work at each place responsible for maintenance when major failures occur.

Fig. 32.1 Some key steps in incident management main processes

In a similar manner, Incident Management is a "normal time" process which can switch into a **Disaster Management** mode when the extent of the incident (or incidents) takes extraordinary dimensions. The boundary between Service Incident management and Network Resource Problem Management in this case disappears, giving an overall "crisis management" cell with close links to the field maintenance organization.

33 Configuration and Change Management

This section covers tasks, processes, and tools necessary for adding, documenting, and modifying network assets, connections, and management tools into the communication system. Depending on different frameworks and vocabularies, it can comprise asset management, configuration management, change management, capacity management, and release management.

In the power utility telecom context, some of the main tasks and processes are as follows:

- Network and Service Inventory
- User Order Handling and Service Activation
- Network Configuration, Capacity Provisioning, and Change Management
- Service Policing and Usage Metering
- Asset Lifecycle and Spare Management
- O&M tools and platform management.

33.1 Configuration Database—Network and Service Inventory

A fundamental element for operating and maintaining telecom systems is accurate, structured and rapidly accessible information on the equipment and connections constituting the network for the various O&M actors. Some of the main O&M tasks requiring such information are the following (Fig. 33.1):

- Problem management tasks require knowledge of equipment type and of the relationships between the alarm generating assets to perform Root cause analysis, as well as assignment coordinates for intervention,
- Determining user service impacts of network faults requires knowledge of service-to-network mappings and dependencies as well as user coordinates for notification,

© Springer International Publishing Switzerland 2017

C. Samitier (ed.), *Utility Communication Networks and Services*,
CIGRE Green Books, DOI 10.1007/978-3-319-40283-3_33

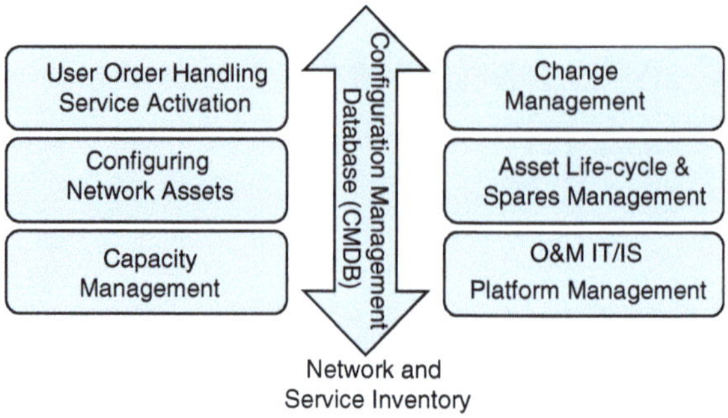

Fig. 33.1 Network and service inventory constitute the core for many O&M activities

- Service restoration and circuit provisioning (i.e., configuration changes) require detailed knowledge of the existing configurations.

The Configuration Management Data Base (CMDB) is the generic name given to the structured presentation of all the concerned information which can include a number of different data inventories:

- **End User Entity inventory** is the list of service users and contact coordinates (phone number, mail address, etc.) for notification and reporting. In the utility operational context, it contains very little information and can be an excel list or included into the service management or integrated management platform.
- **SLA inventory** is the service catalog containing the elements of service quality (availability, time and error performance, required security, outage and restoration time, etc.) for each type of service delivered to users needed for SLA monitoring. In the utility operational context this is again a small volume of information that can be integrated into a performance monitoring or integrated management platform.
- **Service inventory** is the list of all user-referenced connections (e.g., SCADA circuits) and their corresponding attributes such as their mapping to logical resources and connections, SLA elements, configuration, test records, and performance results. This inventory provides service-to-network mapping needed for service impact determination.
- **Logical Resources inventory** is the list (possibly nested) of static and dynamic logical connections across the network providing the network-facing view of the user services: network references (VLAN numbers, etc.), connection capacities and bandwidth, security mechanisms, and underlying physical resources (dependencies).
- **Passive/Active Physical Resources inventory** maintains records of network infrastructure including passive resources such as cabling, cable ducts, etc. It comprises type, version, allocation, configuration data, and status details as well as test and performance results. Resource Inventory is the basis for capacity planning and change management as well as network migration planning. These

applications are used to discover and manage utilization of resources including cable pairs, wiring and distribution panels, and other passive resources.

- **Software/firmware inventory** contains all software releases of tools and platforms as well as all software licenses for the telecom management and service delivery applications used by the O&M organization. This can be part of the IT asset management information systems of the power utility or managed separately by the telecom O&M.
- **Spare parts inventory** keeps track of modules and loose equipment in the utility warehouses, modules sent for repair and expedited orders. Spare management in most cases is a distinct activity of the O&M organization not requiring connections with other management platforms.
- **Supplier/Partner Contracts inventory** is a list of procured services (connections and support services), their pricing, and provider SLAs. In most utility contexts it is a small volume of information not needing any complex data structure.

These inventories can be integrated into a single information system, or maintained as distinct data structures (tables, drawings, etc.) with or without interconnections. Data structures are generally integrated or interconnected for automating processes which are highly recurrent or manipulating very large volumes of data. At the same time, managing interconnected data and maintaining automated processes can considerably increase the complexity of the information system and consequently the cost and effort for its maintenance.

Another issue in adopting strategies on inventory system is the **Inventory Information Model** or the level of detail and resolution of the stored data. The effort needed for populating the information system and keeping it accurate and up-to-date can grow extremely high and the data quality risks to degrade rapidly rendering impossible any automated processes. It is often preferable to have rapid access to less detailed data on the system than having access to detailed but inaccurate information. Detailed information remains always available at its source.

More detailed information is often available in the configuration files of vendor-specific Element Managers (e.g., SDH, and Access equipment). For dynamic logical resources such as virtual connections, the network element itself contains the most up-to-date information which can be read periodically or on-demand into a management platform (network auto-discovery).

In general, we can say that in the operational telecom context, there is a midway solution between no structured inventory and a full-information inventory which may be over-dimensioned and which may require substantial investment to maintain. This midway solution is the upload of "just enough information" needed for limited automations which are to be deployed in their management processes from whatever source where these information reside (e.g., network elements, vendor-specific element managers, excel sheet tables, etc.). The full-scale information shall continue to be available in its original sources. The upload can be made automatic, scheduled, or manual using exchange file import and export.

33.2 User Order Handling and Service Activation

For initializing a new connection of a preexisting type (e.g., a new SCADA connection for a substation RTU using IEC104), or for modifying an existing connection (e.g., increase bandwidth) a possible sequence of steps is given hereafter.

1. *Service Manager Receives User Request (type, allocation, quality attributes)*
2. *Analyze the Request, Check, and Validate feasibility*
3. *Create new SLA with user (if not already existing)*
4. *Register new Service (name, allocation, SLA,* etc.) *in Service Inventory*
5. *Check disruptions and impacts of the demand*
6. *Order Service Initialization/Change*
7. *Coordinate with users, change management, Plan scenario, and* schedule
8. *Notify the User (date of Service Activation)*
9. *Design logical connection (logical connection name, quality attributes, network plane, physical resource usage)*
10. *Register in logical connection Inventory*
11. *Produce Physical Resource Order*
12. *Allocate and Configure physical resources (logical link, bandwidth),*
13. *Test and activate the service over logical connection.*

Initialization or change may be feasible from the network operation support system or may require field intervention. In both cases, the process must include time coordination for performing the requested actions and in the latter case, also site access coordination for the field intervention staff as discussed in a further section.

33.3 Configuration and Change Management, Capacity Management

Configuration management is the overall process of adding, removing, and modifying assets and connections keeping track of all user services, connections, and equipment in a maintained configuration management data base (Inventory treated previously).

Configuration change is not a highly recurrent activity in the power utility. As already mentioned in Fig. 30.6, the creation and remote modification represents a once-per-week (up to once-per-day) action for many utilities.

Configuration change is generally performed using vendor-specific management systems for static allocation communication systems (e.g., SDH, PDH) and potentially off-the-shelf enterprise network tools for dynamic allocated packet networks (e.g., IP addresses, VLANs, Firewall settings, etc.). However, the practice of configuration management and storage of asset configuration files needs to be reconsidered with the new technologies arriving in the network:

- Connecting to the actual network element to obtain settings for equipment is convenient and useful for knowing the current settings in the equipment and is certainly more accurate than previous methods relying on drawings and lists to record settings. However, it can only indicate the current settings and not "what the settings should be from a network design perspective" or in other words "do equipment settings correspond to the intended network design?"
- Each new technology (or the newer generations of the older technology) brings its lot of new functionalities requiring ever increasing configuration settings with significant implications for network operation and performance. Electrical power Utilities may have robust procedures for the design, implementation, testing, and commissioning of such systems to ensure that they perform as required. However, as the systems have become more complex, the trend has been for more settings to be finalized at the commissioning stage. Sometimes, these settings are not recorded in any design documentation and where they are recorded, changes made during commissioning do not always get updated in the original documents. Reliance is placed on the network management system to provide these settings as needed.
- When the network element is replaced, the repairer may need to enter the settings manually and if there is no indication of alarms, extensive regression testing of system operation is not done. The risk is that one or more settings with network implications is incorrectly set and will not be discovered until there is failure elsewhere in the network at a much later time (with potentially serious results).

Such problems may occur with synchronization settings in TDM networks and may also be suspected for security settings in packet switched networks. FTP upload of configuration settings enabled in many recent equipment types associated to an appropriate document management system and process may overcome these issues.

Some of the main tasks to be assured by an enhanced configuration and change management process are as follows:

- Follow up in coordination with field intervention staff, of addition, modification, and suppression of equipment, cable segments and physical links, with storage and documenting of all changes,
- Label all equipment, cables and ducts with unique Identifiers,
- Perform or follow up the creation, modification, and suppression of all logical connections through remote or local access and store in relation with field intervention and network operation support. Document all changes for access of network planning and problem management,
- Label all logical/physical connections with unique Identifiers,
- Keep track of versions, ownerships, settings, factory serial numbers and acceptance tests, manuals, maintenance history, and other asset information,
- Keep track of resource usage and capacity in terms of bandwidth, ports, boards, fibers, and any other shared resources. Examine any new service provisioning

demand for allocating resources subject to resource availability and recover liberated bandwidth when a service is discontinued (Capacity Management),
- Ensure that committed and peak information rates stipulated in the SLA are met (Quality of Service) and limit traffic throughput of Ethernet data ports (Traffic Policing) to avoid the saturation of network planes,
- Establish different projection views of the network infrastructure (e.g., as running today, after the current works, by the end of the on-going project, by the end of the planned transformations, etc.) together with network planning and transformation staff.

33.4 O&M Tools and IT Platform Management

Telecommunication network and service delivery organization increasingly employ IT platforms for their operation and maintenance. These IT platforms also require to be maintained through appropriate process definitions as described in IT governance frameworks such as ITIL. Applied to a telecom service provider, ITIL service management principally defines the processes necessary for the delivery of IT services (IT view) necessary for the fulfillment of the Telecom Provider's Business processes. These include telecom network management system (NMS), trouble ticketing and incident management tools, as well as Voice and data servers for operation related activities.

Some major domains in this context are:

- Software Release and License Management
- Service Support Contracts for constituent Applications and Firmware
- Security and Patch Management
- Software Documentation and IT Configuration Management.

33.5 Asset Lifecycle and Spare Management

An issue of particular importance in the operation and maintenance of an operational telecom network is the increasingly short obsolescence cycle of technologies and communication equipment. The O&M organization must constantly be aware of the situation of different network assets in this respect including commercial availability of equipment for new projects, for replacements, and spare parts as well as support for repair.

Lifecycle management keeps track of the state of all electronic equipment and other elements of infrastructure constituting the power utility telecommunication network (ordered, received, under test, live, under repair, withdrawn, etc.).

Asset Lifecycle management in coordination with different equipment suppliers must anticipate on the end-of-life of equipment and the nonavailability of spare parts constituting stocks as necessary. It must also keep inventory of non-allocated equipment and spares dispersed among different maintenance centers and spare stores.

Another related issue is the determination of the time at which it is no longer economical (even if feasible) to repair or to upgrade assets favoring the replacement of a communication asset with newer generation but functionally equivalent equipment. Naturally, the decreasing cost of telecom technology gradually decreases this age limit resulting in a continual migration of the network.

34 Quality and Performance Monitoring

Quality of communication service and technical performance of telecom systems are closely related concepts often employed in a confused manner.

Quality is the user's perception of an end-to-end service comparing the behavior of the delivered service to the requirements of the application it should serve. It therefore measures the adequacy of a service to an application and as already stipulated in Chap. 19, can be monitored by the application itself (e.g., power system SCADA) or by an external measuring device connected at the Service Access Point. In data networks, Quality of Service (QoS) commonly refers to the capability to serve promptly a given traffic flow through allocated and reserved bandwidth or through giving fair priority hence avoiding queuing delays or loss.

Performance, on the other hand, is a measure of technical parameters defining the proper operation of the system and its meeting of performance objectives. It is generally measured and monitored by the telecom device or less often external measurement tools connected into the system. Telecom equipment in their great majority incorporates monitoring agents which measure periodically several performance parameters which are stored in a management information base (MIB) and can be retrieved by a remote monitoring platform for example through an SNMP GET(*value*).

Performance anomalies detected in the system are to be reported to the incident management for initiating an intervention before impacting the user application and hence avoiding a user-perceived QoS degradation.

Unlike quality monitoring, performance parameters and monitoring techniques are highly dependent upon the nature of the system and are therefore analyzed separately for TDM and for packet networks.

© Springer International Publishing Switzerland 2017

C. Samitier (ed.), *Utility Communication Networks and Services*,
CIGRE Green Books, DOI 10.1007/978-3-319-40283-3_34

34.1 TDM Transmission Performance Monitoring

Performance monitoring for circuit mode digital transmission systems (e.g., SDH) is mainly based on error measurements for both wired (e.g., optical) and wireless (e.g., microwave radio) systems. Monitoring data incorporated into the continuous data stream allow error and continuity measures by incrementing error counters and hence estimating error performance parameters (error second, degraded minutes, etc.). Vendor-specific management platforms are currently used for this monitoring.

Optical TDM systems have practically no errors in normal operation. Their performance is relatively constant unless some fault (e.g., loss of synchronization) occurs, in which case the performance degrades abruptly from fully satisfactory to anomalous and unusable. Performance monitoring is therefore used for fault detection.

In wireless and power line carrier systems, on the other hand, propagation conditions (e.g., atmospheric conditions, fading, noise, etc.) can cause variations of performance which must be monitored.

34.2 Packet-Switched Network Performance Monitoring

Packet network performance depends on the load of the network and on the routing of the packets across the network. It therefore needs to be closely and continuously monitored if time-critical and time-sensitive operational applications are to use the network for their information exchange. The most significant parameters to monitor are packet delay (latency), delay variations, and packet loss.

Moreover, in the case of packet-switched communication there is no continuous traffic flow (unlike TDM) to be used for implicit monitoring of performance. In the absence of any special mechanism, anomalies would only be detected when a data exchange is initiated by the user.

This problem can be overcome by OAM mechanisms, incorporated into the switching equipment, exchanging OAM packets continuously between network nodes. These are defined by IEEE 802.3ah at link-level (i.e., between consecutive switches) and ITU-T Y.1731/IEEE 802.1ag at service level (i.e., end-to-end Ethernet service). These OAM mechanisms have been designed into Carrier-grade Ethernet and reused in MPLS-TP to fill the gap with the existing transport technologies (e.g., SDH) and hence enabling the operator to detect faults and to measure performance. A multilayer approach is adopted like in SDH management (i.e., Regeneration Section Overhead, Multiplex Section Overhead, etc.) and at each layer OAM packets are exchanged between Maintenance Edge Points (MEP).

Among OAM functions, those relating to performance monitoring are

- Loss measurement (LM)—MEPs exchange LM messages every second and calculate loss rate from the number of sent and received messages.
- Delay measurement (DM)—MEPs exchange time-stamped DM messages every second and calculate the delay.

The packet switching nodes maintain performance counters in their MIB which can be collected by a monitoring platform and compressed for report generation over long periods of time.

Telecom O&M Communications and Field Worker Support

35

As mentioned in Part I, remote access to substation-based devices and field workers is an important communication-based application for the power asset maintenance. For maintaining the telecom network itself in operational condition also, remote access from O&M platforms to field device and secure IP access in the substation for field workers are growing applications. The following paragraphs apply equally to telecom O&M and to power system assets maintenance.

35.1 Telecom O&M Communications

Telecom O&M actors require communications to render possible the processes described in this part and some network capacity is hence allocated to their own applications as a network user (refer to Fig. 35.1).

The communication requirements can be broadly classified into

- **Device to platform**—Telecom asset monitoring and fault detection, security barriers reporting to Security Operation Centre (SOC), …
- **Platform to platform**—Connection of vendor-specific management systems and asset inventories to integrated management platforms, User authentication through remote security server (e.g., RADIUS), …
- **Device to human**—Remote access to assets for diagnostics, configuration, and setting (tele-maintenance), Video monitoring of assets and facilities, …
- **Platform to human**—Remote access to telecom management platforms; Automated notification of faults and service impacts to users and O&M work-force, …
- **Human-to-human**—Work assignment, field worker support, O&M reporting, interaction with service users, …

© Springer International Publishing Switzerland 2017

C. Samitier (ed.), *Utility Communication Networks and Services*,
CIGRE Green Books, DOI 10.1007/978-3-319-40283-3_35

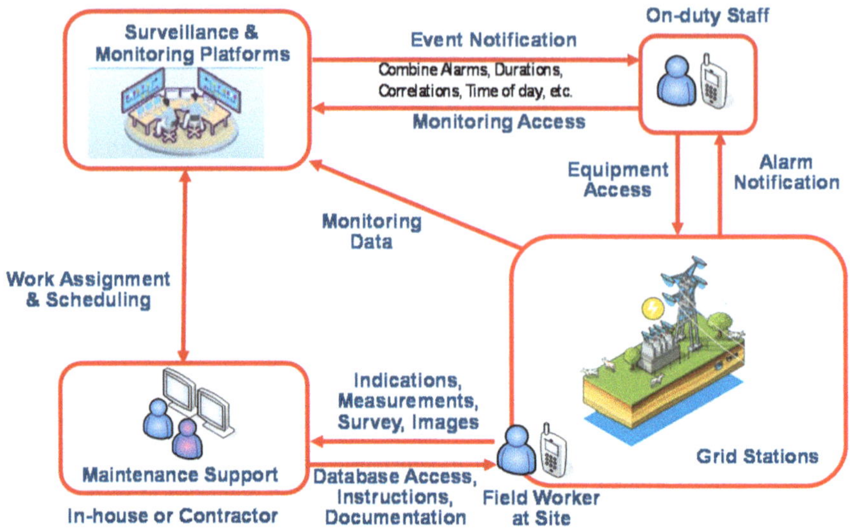

Fig. 35.1 Utility telecom O&M communications (excluding the service user)

IP networking is the most common way for implementing these O&M communications and it is reasonable to constitute a specific IP network for management purpose. Associating remote operators with devices and platforms in secure domain presents important cyber-security issues.

35.2 Connecting to Field Device and Management Platforms

Device-to-Platform communications have mostly been discussed under fault and performance management and employ either embedded channels (e.g., reserved bits or frames in the communication channel) or a dedicated IP connectivity (e.g., SNMP). Many power utilities use encrypted access from field device to management platforms at least partially. This can be performed through SNMPv3 encryption capabilities. However, not all deployed devices support SNMPv3. Moreover, the interface between some communication equipment and their corresponding vendor-specific management platforms is proprietary and not necessarily encrypted. Device-specific tools and platforms can also connect to higher level management platforms through SNMP (north-bound interface) to provide full or partial reporting of events and performances of managed devices.

Connecting an operator's workstation remotely to an O&M platform, to a device-specific tool, or to a field device requires even more attention in terms of security (Fig. 35.2). Some precautions frequently taken in this respect are as follows:

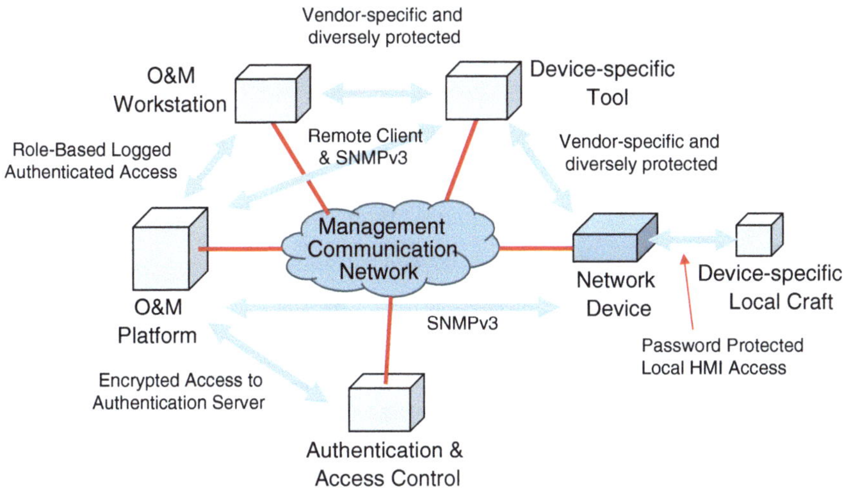

Fig. 35.2 O&M communications around the network device

- Limited access rights for remote operation
- Extremely restricted connection to public network (or no connection)
- Password protection as the minimal security level for connecting to platforms and HMI access
- Centralized access of individual crafts and tools (no direct access from corporate LAN)
- SSH encryption for the access of O&M workstations to management platforms
- Authentication server (RADIUS or TACACS) for remote access to platforms.

35.3 Human-to-Human O&M Communications

Service User to Telecom O&M (Service Manager)
Service desk, which may be highly automated for example in the power utility customer context, is rarely a formal and documented process for the service user calling for internal O&M intervention.

Between O&M actors (field, support, and supervision staff)
Phone and e-mail, are possibly the most common ways different O&M actors interact. Enterprise interworking and specific telecom operation support software applications are mentioned in rare cases.

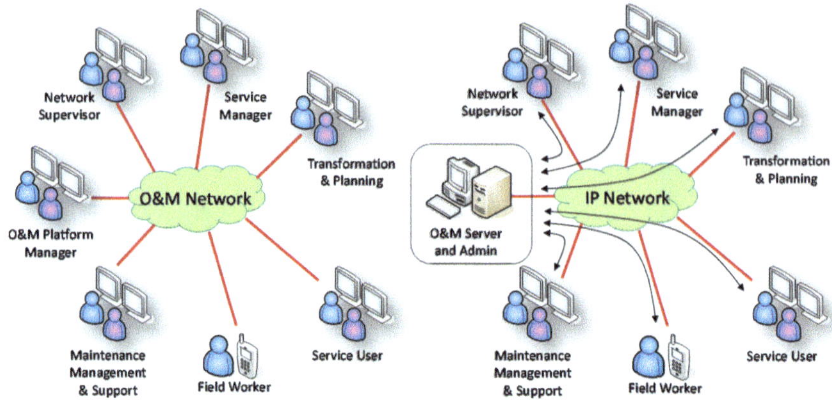

Fig. 35.3 O&M communications between human users

Figure 35.3 presents the main "human actors" that need to be connected:

- Service User to Service Manager for enquiry, reporting, dashboard and notification,
- Field Worker to Maintenance Management & Support for field support and reporting,
- Network supervisor to Maintenance Management for Incident Management, etc.

In an operational context with a small number of actors these communications can be assured directly through an integrated management platform if the actors have access to a client of the management server and hence to profit from the authentication and other security measures implemented into it. The communication can be embedded into the management applications (e.g., incident assignment to a maintenance engineer) or through a dedicated messaging application resident over the management server.

35.4 External O&M Interventions

Some O&M actors may be outside the perimeter of the dedicated telecom network:

- Network supervision may be performed by an external contractor
- Field worker may be accessible only through a public mobile network
- An external supplier's or service contractor's field worker at site may need to access his support platform located outside the operational network.

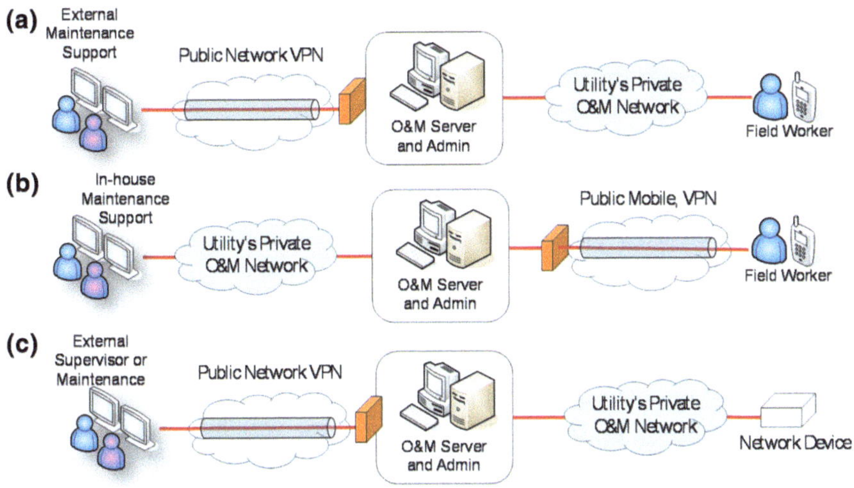

Fig. 35.4 O&M actors beyond the perimeter of the private dedicated operational network

In this case, all or part of the communication needs to pass over a public communication network and supplementary security measures such as the usage of VPN isolation (e.g., IPsec) need to be considered in addition to centralized authentication. Figure 35.4 illustrates these particular cases.

35.5 Field Worker Access to Operational Sites and Assets

Field intervention is the most demanding part of the operational telecom O&M process requiring a relatively important workforce and logistics depending upon the size and geographical dispersion of the network and on the required intervention time. It can typically take from few hours up to several days depending upon availability of workforce and severity of impacts and consequently the priority.

Power utilities mitigate this issue through

- Remote intervention as far as possible, for diagnostics, reconfiguration, and for service recovery by avoidance of faulty device, easing hence the urgency of site intervention limited to module change operations and advanced diagnostics. Remote access to field device has already been discussed in previous sections.
- Power utility's multiskill field intervention workforce or external contractors. Nonspecialist field intervention requires remote specialist support from in-house or external maintenance support platforms.

Physical Access to Field Sites

Access to unmanned electrical power substations is a problem for field intervention generally resulting in an increase of intervention time. Different strategies can be adopted as follows:

- Same access key for all sites and in possession of all intervening parties—presents no security and no access control
- Differentiated access keys for sites at maintenance centers made available to site intervention workforce—presents little control on site access and no flexibility
- Remote access control through phone request at site entrance—Authentication remains basic and not very secure but simple to implement
- Electronic key (card-based access control) with centralized authentication and authorization enabling access certificates when required through remote request—can be a solution providing fair security without very high complexity. The solution, however, is not disaster-proof (needs power and communication to operate) unless backed up with a local fallback process and dedicated long-life battery power.

An association of different strategies can be employed for an economic, flexible, and secure solution.

Local Access to Assets inside Field Sites

Field worker's local access to the Human–Machine Interface (HMI) of different assets (whether communication equipment or substation automation device) is another issue which needs to be solved. The field site itself is often considered as a physically secure environment meaning that only authorized persons are supposed to have access to the perimeter of the field site where the assets are located. This can be further secured by partitioning of the field site: an electrical substation can for example have separated physical zones for differentiated access to assets for different types of intervention. Cables, connection sockets, and equipment HMI ports can also be made secure by using appropriately locked conduits and cabinets.

On the other hand, there is an increasingly frequent demand to explore the provision of central or wireless access for the intervention staff inside field sites. The objective may be to avoid physical connection and disconnections to a great number of HMI ports, some of which being hardly accessible and forcing the field staff to check status, to perform measurements and to change settings in unstable, uncomfortable and sometimes unsafe positions. Substation HMI access servers may be used to provide access based on user credentials to a selection of field asset HMI. It can further provide local (or remote) log management information as well as tagging, work permit checking, local schematics and configuration management data. Secured WiFi is used by some utilities for local access to HMI (not the operational LAN) and is being incorporated as an option to many telecom assets.

Remote Platform Access from Field Sites

Field worker's communication to central facilities and remote expert support inside and outside the utility infrastructure is a particularly important issue for both internal and external interventions at site.

Figure 35.4a represents the connection of a field worker to an external maintenance support center using the utility network for connection to an O&M platform which serves as an access point through a public VPN. For an internal maintenance support center, the connection from the O&M platform is directly through the operational network. Dedicated servers and separate LANs are sometimes deployed at field sites for connecting field worker PCs in a corporate LAN and hence isolating them from critical applications.

Figure 35.4b shows the usage of public mobile network for connecting the internal or external field worker to a maintenance support platform. The connection must again be done through a VPN and an appropriate security barrier must be placed at the intermediate O&M platform. Many Ethernet switches provide mobile data service connection with IPsec VPN service isolation.

Isolated Worker Safety Communications

Health and safety regulating authorities in many countries provide particular rules to deal with the case of workers whose duties bring them to work alone in a site, in order to protect them from the consequent hazards. Emergency situations may arise due to sudden onset of a medical condition, accidental work-related injury or death, attack by an animal, exposure to elements, or by becoming stranded without transport, food or water. A person is considered as alone at work when he cannot be seen or heard by other persons who can provide assistance if necessary.

The power utility must ensure that a means of communication is available in the event of an emergency to enable the employee to call for help, and also ensure that a procedure for regular and systematic contact with the employee at predetermined intervals has been established.

1. Fixed Telephone service—The simplest communication service for isolated workers is indeed the accessibility of a telephone at site, provided that the person is able to reach the telephone in an emergency. This service must be available not only to the operational staff but also to external parties contracted for specific tasks in the utility premises. The telephone system must provide an emergency number accessible to all categories of users.
2. Radio communications—Different categories of two-way radio systems are in use in different Utilities for traveling staff and for employees working in large sites such as power plants. Mobile workforce communications were covered in part I, Sect. 6.2 and may also apply for assuring the safety during the trip to site depending upon coverage constraints. Usage of private mobile radio as a means of assuring worker's safety needs careful location of base stations and identification of shadow areas as well as an adequate procedure to assure the supply of charged batteries. Public mobile phone is used in some Utilities as a cost-saving alternative, but may present serious drawbacks with service coverage and

service continuity particularly when the power utility employees are intervening to reestablish electrical power in a region due to the extremely short power autonomy of the public base stations.

3. Satellite Communication systems—Satellite phones overcome the problems of public mobile phone in poorly covered areas and the dependence on local power supply of the base stations. Satellite systems also allow implementing "Location Beacon Systems" determining the location of the employee through GPS and signaling this location to an operational base permanently. Care should be taken, however, as their operation is affected by damage to aerials, failure of vehicle power supplies, or vehicle damage.
4. Personal security systems—These portable wireless transmitters are permanently in communication with a central receiver and may include a non-movement sensor that will automatically activate an alarm transmission if the transceiver has not moved within a certain time.
5. Emergency location beacons—When working in particularly remote areas, emergency location beacon systems which are automatically activated in emergency situations may be used. These systems do not depend upon the vehicle power supply and do not risk damage as satellite communication systems.

35.6 Disaster-Mode Operation

As already stipulated in the introductory paragraph of this part, disaster-mode operation consists in managing major incidents with "disastrous" impact on the service in order to maintain service continuity for the most critical services using special processes, and through mobilization of extra people and extra tools as compared to those employed for normal operation. The target is no longer optimality but rather the survival of essential services. It is clear that the power utility cannot afford being in disaster-mode very frequently and the decision to "declare the disaster situation" must remain an exception. On the other hand, remaining in normal operation under an exceptional situation can have serious consequences for the utility, for its staff and for its customers. It is therefore important to have clear indications of the condition for declaring disaster mode and the authority making the decision.

Disaster-mode operation signifies extra people and extra tools from other regions or from other activities, postponement of some normal recurrent activities and a reduced process. Typically incident management, service enquiry and problem management can be squeezed into a single crisis cell with permanent and reinforced communication with an enhanced field intervention workforce. The telecommunication network may be totally or partially out of use. Fast deployment telecommunication infrastructure described in Part IV, Sect. 27.3 is therefore an important part of the solution. Private mobile radio and satellite communication bandwidth are relevant solutions in this context. Rapid estimation of usable communication resources and consequent reallocation of resources allow building the necessary communication infrastructure. Precise knowledge of service priorities is a key issue here. The following are some measures towards an enhanced disaster recovery plan:

- Train personnel with emergency procedures
- Notify key personnel of the problem and assign them tasks focused toward the recovery plan
- Notifying clients about problems minimizes panic
- Identify and assure access to tools needed for the recovery (manuals, procedures, privileges, etc.)
- Backup data should be stored in separate locations
- Backup sites and mobile recovery facilities
- Recovery plans must be tested frequently and updated continually
- Prepare employees for working under stress and a support system to alleviate some of the stress
- A disaster information system allows the utility to acquire accurate information across the network in order to mitigate damage from disasters
- Improve contingency plan for power communication under various categories and at various levels
- Set up special maintenance talent teams for fast response to reduce business interruption time.

Appendix 1
Termination Networks and Service Access

Terminating the telecommunication network at Service Access Points (SAPs) where interfaces are delivered to the User Applications (or their LAN gateways) merits further discussion. This part, whose subject merits a book on itself, is produced as an appendix in order to avoid breaking the overall coherence.

The termination segment of the network can employ many dedicated technologies, due to the specific constraints of the end sites, and may require specific considerations on site cabling, installation practice, and electromagnetic compatibility issues. Here we focus on high-speed digital interface and in particular on Ethernet as the increasingly "universal" service access point in power utility services.

Moving from individual wiring of applications through analog or low-speed data access to Ethernet reduces the immunity to fast disturbances and transients which can no longer be filtered out as easily as with low-speed signals. This means that increased precaution is necessary to avoid capturing disturbances (through balanced pairs, screening, grounding and shielding). Optical fibers may also be adopted with low-speed signals in harsh local area to provide electromagnetic immunity.

The maximal distance across which the interface can be extended in the local area is also greatly impacted by the move from low-speed wiring to Ethernet. If in the former case, the limiting factor was essentially "loop resistance," cable attenuation and signal distortion due to inter-symbol interference limit the maximal span of the signal in Ethernet. Again the usage of optical fibers in the local area access may provide superior span capability for high-speed data transmission. A major drawback of optical fibers in the local area is however the power supply wiring: Ethernet over copper wire can transport electrical power for the end devices while dedicated wire is necessary for local optical networks.

Passive optical networks transporting Ethernet (EPON) is a potential solution in the approach network in smart distribution environment. However, underground copper cables are still abundantly present and may be used together with high-speed Digital Subscriber Line (xDSL) techniques.

Other technologies such as Power line communications and Broadband wireless packet already described in Sects. 22.9 and 22.10 can also constitute potential solutions for bringing network connections to SAPs in customer premises and to distribution substations and assets.

© Springer International Publishing Switzerland 2017

C. Samitier (ed.), *Utility Communication Networks and Services*,
CIGRE Green Books, DOI 10.1007/978-3-319-40283-3

Table A.1 Ethernet approach network technologies

Domain of use		Span	Transmission technology
Equipment room	Cabinet level (backplane)	Few meters	Copper wire
	Local interface to application		Copper or fiber interface
	Temporary and nomadic access		WLAN (IEEE802.11)
Substation control building	Automation LAN Video-surveillance and access security LAN	10–100 m	Fiber interface
	Voice and data LAN		Fiber, wireless (IEEE802.11)
Multiple substations and buildings on one site		100–1000 m	Fiber Ethernet Wireless Ethernet link DSL copper access
Small geographical region	Urban interconnections Energy farms, Hydro plants	1–20 km	GbE over fiber Broadband wireless mesh (IEEE802.11)
Remote access to electrical sites	S/S on fiber network	1–x km	Ethernet over SDH Ethernet over Fiber
	S/S not on fiber network		HV PLC, MV BPL Wireless
Remote access to device on customer premises	Advanced metering infrastructure	1–10 km	DSL access BPL (IEEE 1901) Cellular 2G/3G/4G Broadband wireless Ethernet PON NB PLC (PRIME, etc.)
Remote access to facilities at operation support sites	Site on fiber network	5–50 km	Ethernet over SDH
	Site not on fiber network		Microwave Ethernet, IEEE802.16 VSAT Carrier Ethernet service

The mentioned technologies (as well as some others) constitute a panel of signal transmission solutions that often need to be combined for responding to different situations and constraints encountered in the termination networks of electrical power utilities.

Table A.1 presents some Ethernet approach situations and corresponding technological solutions.

Optical Fiber in the Termination and Access Network

Optical cables used for connecting Ethernet nodes and devices employ the following types of fiber:

- Graded Index Multi Mode (GI MM) fiber (62.5/125 or 50/125 μm)—These fibers are mainly used inside buildings for LAN applications.
- Single Mode (SM) fiber with a core diameter of 8–10 μm (known as 9/125 or 8/125 μm) is the traditional fiber used for long-distance high-speed communication.

Multimode Fiber (MM)

Optical installations inside buildings in their large majority employ multimode 50/125 μm fibers (ITU-T G.651.1) although 62.5/125 μm fibers may be encountered in some old installations. Multimode fiber is still popular because it can be associated with less costly and less precision-sensitive LED/PIN transceiver technology and connectors in particular for Fast Ethernet (100 Mbps). If Laser light source and APD receivers are to be used, then multimode fibers present little benefit and severe performance limitations due to modal dispersion. Special attention is therefore necessary when upgrading the speed and transceivers on existing indoor optical cabling. Table A.2 presents some typical values with LED/PIN transceivers

Table A.2 Multimode optical fiber standardized performances (vendor values in brackets)

ISO/IEC 11801 Fiber Type	Wavelength (nm)	Bandwidth (MHz/km) IEC (typical)	Interface	Max. distance (m)
OM-1 (OM-1+) 62.5/125 μm	850	200 (200)	1000 BASE-SX	275 (500)
	1300	500 (600)	1000 BASE-LX	550 (1000)
	850	200 (200)	10 GBASE-SR	33 (65)
	1300	500 (600)	10 GBASE-LX4	300 (450)
OM-2 (OM-2+) 50/125 μm	850	500 (600)	1000 BASE-SX	550 (750)
	1300	500 (1200)	1000 BASE-LX	550 (2000)
	850	500 (600)	10 GBASE-SR	82 (110)
	1300	500 (1200)	10 GBASE-LX4	300 (900)
OM-3 50/125 μm	850	1500 (2000)	1000 BASE-SX	860 (1100)
	1300	500 (500)	1000 BASE-LX	550 (550)
	850	1500 (2000)	10 GBASE-SR	270 (300)
	1300	500 (500)	10 GBASE-LX4	300 (300)
(OM-3+) 50/125 μm	850	(4700)	1000 BASE-SX	(1100)
	1300	(500)	1000 BASE-LX	(550)
	850	(4700)	10 GBASE-SR	(550)
	1300	(500)	10 GBASE-LX4	(300)

according to ISO/IEC 11801. For new fiber installations where GbE may be involved, the installation of SM fiber or a mix of SM and MM fiber should be considered.

Single Mode Fiber (SM)

Single mode 9/125 μm fiber is the traditional long-distance high-speed telecommunication fiber. The currently used type is ITU-T G.652 with a zero-dispersion at 1324 nm. Different categories (A, B, C and D) are given in the standard with different levels of performance. In the inter-site or intra-site connections, it is associated with 1310 or 1550 nm Laser transceivers, and may be used at multiple wavelengths through DWDM/CWDM (types C and D). The zero-dispersion point of the fiber can be shifted through adequate doping for extending the span at 1550 nm (ITU-T G.653) but with degraded performance outside this wavelength which makes the fiber unusable for Wavelength Division Multiplexing. A "near-zero dispersion shift" fiber standard ITU-T G.655 allows the usage of WDM over extended span lengths.

Optical Connectors

Many different optical connector technologies are available on the market including some application-specific types and some which are no longer used into new deployments. Here, only those types which are commonly used, in particular in harsh environments are discussed.

Some particularly relevant aspects when selecting connector types in an industrial environment, are tensile load (mechanical resistance), durability (number of mating cycles) and operating temperature, in addition to transmission characteristics such as return loss (back reflection) and attenuation. Packing density is also a major aspect when a great number of connections have to be made over a distribution panel (Optical distribution frame or ODF).

Most connectors employ a mating adapter which may be for MM fibers, for SM fibers, or for both.

Similarly, different polishing methods are employed in connectors which can be categorized into flat, convex, and angle polished connectors (Table A.3).

IEC 61754 standard provides the functional and transmission characteristics for many different types of optical connectors. Commonly used connectors are SC (IEC 61754-4), ST (IEC 61754-2), and LC (IEC 61754-20). Other types such as FC and SMA are common in older installations.

Table A.3 Different polishing methods used in optical connectors

	Flat polishing	Convex polishing PC: physical contact UPC: Ultra-PC	Angle polishing APC: angled PC
Fiber type	MM	MM, SM	SM
Back reflection (cause interference for laser transmitters)	High return loss (20–30 dB)	Low return loss PC: 45 dB UPC : 55 dB Must be connected	Low return loss APC: 65 – 85 dB Need not be connected

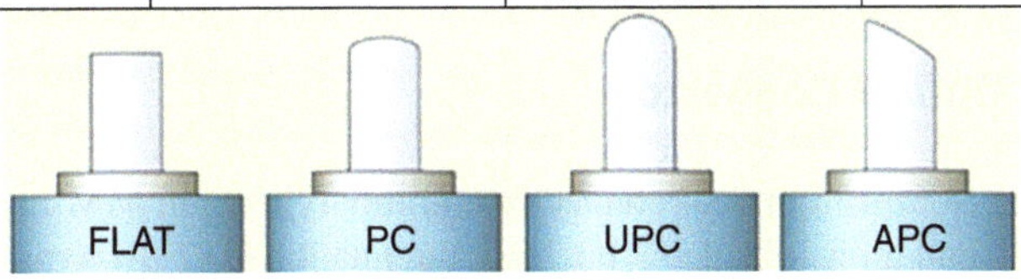

SC-Type Optical Connector

Optical connector type SC is the most commonly used connector in all types of installations with SM and MM fibers. SC connector is of "snap-in" type and is available as single or as duplex. In a 19″ ODF (Optical Distribution Field) up to 48 (24 pair) can be installed. SC has a 2.5 mm diameter ferrule in zirconium with plastic housing. Attenuation is typically 0.2 dB with SM fiber and 0.3 dB for MM fiber.

LC-Type Optical Connector

LC is a newer connector than SC. It has similar optical performance and is available as single or duplex. LC has 1.25 mm diameter ferrule in zirconium with plastic housing. The only advantage is the small size which makes it possible for higher package density connectors in the ODF. LC connectors are used for SM and MM installations.

ST-Type Optical Connector

ST connectors are "bayonet type" connectors for SM and MM fibers commonly used on WAN/LAN equipment. It provides lower package density (due to bayonet fixing) allowing at most 12 connectors in a 19″ ODF, and therefore less common in network installations. When using similar 2.5 mm diameter ferrule, ST has similar performance to type SC.

FC-Type Optical Connector

Optical connector type FC was one of the first high-performance connectors on the market and is used in telecommunications installations all over the world. This connector has screw-thread, and maximum 12 connectors can be installed in one 19″ ODF. FC is still very common in older installations and on test equipment. Using similar 2.5 mm diameter ferrule, FC has similar performance to type SC.

SMA-Type Optical Connector

SMA or SMA 905 is an optical connector commonly used in MM installations up to mid-1990s and now obsolete. Its high attenuation (0.7 dB typical) and low performance need to be taken into consideration when older optical installations are to be reused for new service access requirements. This connector system has screw-thread fixing and maximum 12 connectors can be installed in one 19″ ODF. The connector ferrule has a diameter of 3.17 mm and is often in zirconium although full metal is also common.

Copper Cables in the Termination and Access Network

Copper wire is widely used at the service access point and in the termination network. Some major drivers for the use of copper are:

- Easy mechanical connection using screw terminal blocks, RJ45 or coaxial connectors
- Possibility of remote power feeding
- Extensive availability in industrial plants of legacy wires (multipair cables)

However, copper wires are subject to electromagnetic interference and bandwidth/span limitations requiring particular attention during refurbishment works. As a rule, optical fibers should be recommended whenever an Ethernet communication link or network is to run across an electrical environment such as a substation or a power plant.

Coaxial cables present controlled impedance at higher frequencies (often 50 or 75 ohm) due to their specific geometry and therefore allow the transport of RF communication signals with minimal loss and distortion. However, the unbalanced signal transmission makes them less immune to electromagnetic disturbances. They are at present commonly used for transceiver to antenna connections in wireless systems where the spurious disturbance can more easily be filtered out and the high carrier frequency yields a narrow relative bandwidth. Symmetrical wire pair also known as balanced pair (or twisted pair) is the most current usage of copper wires in the termination network. The impedance at high frequencies is not well controlled (often taken in practice as 100–150 ohm but variable in reality). The balanced

Table A.4 Symmetrical pair copper cable categories

Type	Connector	Frequency range (MHz)	Guidance
Cat 3	8P8C (RJ45)	0…16	This type is not recommended anymore (TIA/EIA-568-B)
Cat 5	8P8C (RJ45)	0…100	Category 5 cable can be used for 10/100 Mbit/s Ethernet Diameter: 0.5 mm (24 AWG)
Cat 5e	8P8C (RJ45)	0…100	Category 5e cable is an enhanced version of Cat 5 to improve the performance with 1 Gbit/s Ethernet
Cat 6	8P8C (RJ45)	0…250	Category 6 can be used with 10 Gbit/s Ethernet over shorter distances (10 m). (TIA/EIA-568-B)
Cat 6a	8P8C (RJ45)	0…500	Category 6a allows up to 10 Gbit/s Ethernet with a cable length up to 100 m. (ANSI/TIA/EIA-568-B.2-10)

nature of signal transmission reduce significantly noise and disturbance due to almost identical coupling to each conductor and consequent common mode rejection for the differential signal.

Main cabling standards in use are as follows:

- EIA/TIA 568—North American cabling standard
- ISO/IEC 11801—International standard for structured cabling systems
- CENELEC EN 50173—European cabling standard

The following table defines different cable categories (as defined by EIA/TIA 568) (Table A.4).

Electromagnetic Compatibility (EMC)

Electrical substations are severe electromagnetic environments, and various types of disturbances can be easily encountered. All electronic equipment installed in the HV electrical substation must be able to resist to electromagnetic phenomena present in this environment. When selecting Ethernet termination devices for use in electrical substations, the communications designer must be aware of the special requirements imposed by the application.

- Conducted and radiated noise—These disturbances are to be mitigated by means of shielding and filtering detailed separately in the following paragraphs.
- Overvoltages—Usually substation power supplies are derived from battery arrays, which may experience overvoltages when charging. Also if the supply is directly obtained from the grid, fluctuations may occur, mainly during disconnections and reconnections in the substation bars. This may lead to equipment breakdown. Substation-grade Ethernet devices must therefore withstand a higher than usual supply voltage range.

- Reverse polarity—Every device shall withstand an accidental reversal of the power supply terminals during installation.
- Electrostatic discharge—The networking equipment shall be able to work under the presence of electrostatic discharges in exposed metallic parts and interfaces. The device must not only avoid damage, but also malfunctioning during discharge. Electrostatic discharge may occur with human contact due to poor electric contact to ground (e.g., inadequate shoes) with certain air conditions (lack of air moisture) resulting in a large difference of potential between the operator's body and networking devices.
- Surge protection—Ground faults and switchgear operations are substation events which originate very intense current flows and discharges. These currents are coupled into any metallic conductor resulting in huge voltages in the power supply terminals of every network device. The exposed devices must have the capability to resist these surges. Common design practice includes voltage spark gaps, safety capacitors, gas discharge tubes, varistors, Zener diodes, etc.
- Failure mode—Networking equipment must be designed to guarantee a benign failure mode. This will guarantee that the failure of a switch power supply, for instance, will not affect any of the devices powered by the same power rails, or even the primary power supply.
- Ethernet links should be on optical fibers as far as possible to provide immunity to harmful electromagnetic disturbances.

IEC61850 in its part 3, "General Requirements" specifies the climatic and electromagnetic environment necessary for Ethernet hardware to be used in the substation. Depending upon the nature of the test, the equipment must either continue to function correctly under the test conditions, or resume normal operation after the condition has disappeared without the requirement for a Reset.

A comprehensive set of immunity tests is given by IEC 61000-4-x standards referenced in Appendix 3.

RF Interference

Every cable or device capable of conducting electricity can potentially pick up radio frequency interference from the surrounding environment. This problem is of great importance in environments where considerable electromagnetic disturbance is present due to large current flows or voltage differences, as in electrical substations.

Certain frequencies are particularly harmful for twisted pair-based Ethernet communications, such as 12.5 MHz, 125 MHz and 1.25 GHz, which are the signaling rates for Ethernet running at 10, 100 and 1000 Mbps.

Vulnerability to RF Interference depends also on the way cabling is performed through the substation, on the employed ducts, on the type of deployed cables, and on the proximity to noise sources (coils, motors, circuit breakers, switchgears). Field-to-cable coupling is not the only type of RF interference mechanism which may be present. Crosstalk between conductors running in parallel for some distance is also a common problem with unshielded cables.

Shielding and Grounding

Shielding is an effective way of reducing the electrical noise and transients coupled into electric cables. It can be very efficient for canceling the effect of electric and high frequency magnetic fields (>100 kHz). However it brings little benefit against low frequency magnetic fields.

There are two main points of interest regarding shielding practices:

Shielding of Cables and Connectors

Electrical cables and connectors are often shielded to protect them from incoming radio frequency interferences. Electrical cables running on long distances in electromagnetically noisy environments are particularly sensitive to these hazards, as they can act as receiving antennas for the incoming radiation.

Shielded cables end in shielded connectors, which, in a typical installation, are electrically connected to the casing, itself connected to the chassis ground. A user may inadvertently create a ground loop by connecting both ends of the cable to chassis ground, separated by tens or hundreds of meters. In fact electrical cables are the main mechanism for coupling noise into electronic equipment.

In Ethernet networks three type of shielded cables are used for 10/100BaseTx or gigabit applications:

Shielded Twisted Pair, STP—every copper pair is individually shielded.
Screened/Foiled Twisted Pair, S/FTP—the four twisted pairs are foiled together with a metal sheet.
Screened/ Shielded Twisted Pair, S/STP—every copper pair is individually shielded, and then the aggregated four pairs are covered by a metal sheet. This type of cable provides the best protection against external interference coupling and crosstalk between pairs.

Shielding of Equipment

Electronic equipment is usually installed in grounded metallic cabinets.

The metal housing can effectively mitigate the electromagnetic field that surrounds the equipment. The electric currents in the metal frame decay exponentially with the width of the material. The skin depth determines the required thickness of metallic screen to attenuate the electromagnetic wave to a certain extent. The skin depth depends on the employed material and decreases with increasing frequency. For instance at 10 kHz, iron has a skin depth of 0.1 mm, whereas Aluminum's skin depth is close to 1 mm.

It should be noted that every slot in the case may act as a receiving or transmitting antenna, since surface currents may flow in the conductor around the slots creating a radiating field.

Power over Ethernet (PoE)

One important aspect that pleads in favor of copper wire connections in Ethernet environment is the possibility to provide remote power supply through the LAN cabling to some network devices. This in particular applies to smaller devices requiring relatively small amount of power:

- Monitoring and surveillance video cameras,
- IP telephone devices,
- Wireless LAN access points
- Site Access Control ID card readers
- Ethernet Microwave transceiver, etc.

Many industrial Ethernet switches provide PoE capability on their copper wire interface ports and hence reduce the necessary site cabling and increase the flexibility of the installations (e.g. Wireless LAN base stations can be repositioned without modifying power cabling).

Furthermore, the use of PoE enables the remote management of power delivery to the concerned devices using the previously described SNMP switch management facilities.

The standard IEEE 802.3af allows the delivery of 48Vdc supply up to around 15 W per port using a Cat5 or Cat 5e copper cable. An enhanced standard, IEEE802.3at allows delivering up to 25 W of supply power.

Two options are available for the operation mode of the Power over Ethernet.

- Power can be injected by the Power Sourcing Equipment (Ethernet Switch) into spare copper wire pairs in the Ethernet cable and recovered into a DC/DC converter at the Powered Device.
- Power can be injected by the Power Sourcing Equipment across the center taps of the two isolation transformers (associated to Transmit and Receive directions) and recovered across the corresponding transformer center taps at the Powered Device DC/DC converter.

A "discovery process" at the Ethernet Switch allows to examine the Ethernet cable and to sense PoE devices with a smaller voltage (2.7–10 V) and limited current before applying the full 48Vdc remote power.

Appendix 2
ITIL Management Framework

Despite the fact that IT management, in general, is beyond the scope of the present volume, it may be useful to provide an indicative description of this framework. The TMF Telecom Operation Map e-TOM and its derived Utility Telecom Operation Map uTOM described in sections 17 and further in section 24 provide a process model essentially for a Telecom Service Provider and describe how telecom services can be delivered to the "customers" of the service provider. ITIL, on the other hand, provides a formal framework for building internal processes in any Information-oriented organization hence including the utility telecom service provider.

The IT infrastructure library (ITIL) gives a detailed description of best practices, provides checklists, tasks and procedures that may be used by IT organizations according to their needs. Through its scalable and flexible "adopt and adapt" approach, ITIL is applicable to all IT organizations irrespective of their size or the technology in use.

ITIL is not a standard, but rather a set of guidelines for a structured, common-sense, process-driven approach to ensure close alignment between IT and business processes. It recognizes that there is no universal solution to the process design and implementation for the management and delivery of IT services. As such, organizations, management systems and tools cannot be "ITIL compliant", but may implement management processes assessed through ITIL guidance.

The ITIL V3 consists of 26 processes and functions grouped under 5 volumes, arranged around the concept of Service Lifecycle structure (given in table hereafter):

- Service Strategy
- Service Design
- Service Transition
- Service Operation
- Continual Service Improvement

ITIL V2 (2001), in particular its first two components (service support and delivery) are the most commonly used. The V2 grouped process guidelines

© Springer International Publishing Switzerland 2017

C. Samitier (ed.), *Utility Communication Networks and Services*,
CIGRE Green Books, DOI 10.1007/978-3-319-40283-3

according to different aspects of IT management, applications and services into 8 logical sets:

- Service Support
- Service Delivery
- ICT Infrastructure Management
- Security Management
- Business Perspective
- Application Management
- Software Asset Management

The logical sets have been complemented by two implementation guidelines:

- Planning to Implement Service Management
- ITIL Small-scale Implementation (for smaller IT units)

The structures of V2 and V3 are given in tables hereafter.

ITIL V2 processes (note that explanations are only indicative)

Service support	Service desk/service request management (SD)	• Provide a single point of contact for service users and handle incidents, problems and questions • Perform lifecycle management for service requests • Keeping the customer informed of progress and advise on workarounds • Handle large volumes of telephone call transactions (call center) • Manage, coordinate, and resolve incidents as quickly as possible at primary support level (help desk) • Provide an interface for other activities (e.g. change requests, maintenance contracts, software licenses, SLM, CM, AM, FM, and ITSC)
	Incident management (IM)	• Restore normal service operation (within SLA) as quickly as possible with the least possible impact on either the business or the user, at a cost-effective price • An 'Incident' is any event which is not part of the standard operation of the service and which causes, or may cause, an interruption or a reduction of the quality of the service
	Problem management (PM)	• Resolve the root causes of incidents • Minimize the adverse impact of incidents and problems caused by errors within the IT infrastructure • Prevent recurrence of incidents related to these errors

(continued)

(continued)

		• A "problem" is an unknown underlying cause of one or more incidents, and a "known error" is a problem that is successfully diagnosed and for which either a work-around or a permanent resolution has been identified
	Change management (ChM)	• Ensure that standardized methods and procedures are employed for handling all changes (add, modify, or remove) • Ensure minimal disruption of services • A change request (CR) is sent to the ChM and reflected into a Forward Schedule of Changes (FSC)
	Release management (RM)	• Ensure the availability of licensed, tested, and version-certified software and hardware, functioning as intended when introduced into existing infrastructure
	Configuration management (CM)	• Track all of IT assets through the CM database (CMDB) containing assets, their configurations, and their interactions • Configuration planning and regular planning reviews • Identification and labeling of assets. Recording of asset information, hard/software versions, ownership, documentation and other identifiers with a business defined level of detail • Control—Liaise with ChM to ensure that no Configuration Item is added, modified, replaced or removed without approved request for change, etc. • Lifecycle monitoring (ordered, received, under test, live, under repair, withdrawn, etc.) • Verification reviews and audits (physical existence, correct records in the CMDB and parts list). Check release documentation before changes are made to the live environment
Service delivery	Service level management (SLM)	• Primary interface with the "customer" (as opposed to the "user" serviced by the service desk) • Produce and maintain a service catalogue (standard IT SLAs) • Monitor the IT service levels specified in the SLAs, ensure that the agreed services are delivered • Establish metrics and monitor against benchmark • Liaise with AM, CaM, IM and PM to ensure required service level and quality • Ensure operational level agreements (OLAs) with support providers

(continued)

(continued)

	Capacity management (CaM)	Match IT resources to business demands • Application sizing • Workload management • Demand management • Modeling • Capacity planning • Resource management • Performance management
	IT service continuity management (ITSC)	• Perform risk assessment and reduce disaster risks • Ensure service recovery and evaluate recovery options • Business continuity planning • Prioritize service recovery through business impact analysis (BIA) • Produce contingency plan • Regularly test and review the plan
	Availability management (AM)	• Survey availability requirements • Produce availability plan • Monitor availability obligations • Manage resilience
	Financial management (FM)	• Calculate and optimize service cost • Recover costs from users
ICT infrastructure management	ICT design and planning	• Strategies, policies and plans • Overall and management architecture • Requirement specs and tendering • Business cases
	ICT deployment management	Design, build, test and deploy projects
	ICT operations	• Day-to-day technical supervision • Incidents reported by users or Events generated by the infrastructure • Often work closely with Incident management and service desk • Logging of all operational events • Maintenance of operational monitoring and management tools
	ICT technical support	Support infrastructure and service management with multiple levels of technical expertise
Security management	Deploy security policy in the management organization	
Business perspective	• Business continuity management • Transforming business practice, partnerships and outsourcing	
Application management	• Improve the overall quality of software development and support • Gather requirements for meeting business objectives	
Software asset management (SAM)	• Maintain software license compliance • Track inventory and software asset use • Maintain policies and procedures on software asset lifecycle	

ITIL V3—"service life cycle"-oriented and network-centric

Service strategy	Service portfolio management	Strategic thinking on how the portfolio should be developed in future
	Demand management	Understand and influence customer demands
	IT financial management	Idem V2
Service design	Service catalogue management	V2: part of SLM
	Service level management	Essentially same as V2 Service review now in CSI
	Risk management	Dispersed in several processes V3: Coordinated process
	Capacity management	Idem V2
	Availability management	Idem V2
	IT service continuity management	Idem V2
	Information security management	V3: Improved integration across service lifecycle
	Compliance management	V2: Addressed within several processes
	IT Architecture management	V2: Covered within ICT design and planning
	Supplier management	V2: Covered within ICT Infrastructure management
Service transition	Service asset and configuration management	V2: Configuration management
	Service validation and testing	V2: Release management extended
	Transition planning and support	• Plan and coordinate resources to ensure that service strategy encoded into service design are realized in service operations • Identify, manage and control risks of failure and disruption across transition activities
	Evaluation	• Ensure that service will be useful to the business. Set metrics and measurement to ensure continued relevance of services
	Release and deployment management	V2: Release management extended
	Change management	Essentially same as V2
	Knowledge management	New process in V3, previously included to some extent in problem management

(continued)

(continued)

Service operation	Event management	Part of Infrastructure management in V2, has been extended as the trigger for Incident and Problem management
	Incident management	Essentially same as V2
	Problem management	Essentially same as V2
	Request fulfillment	New in V3. In V2, service requests were treated by Incident management
	Access management	New in V3. In V2, it was part of security management
Continual service improvement (CSI)	Service level management	New in V3. In V2, this was treated in SLM
	Service measurement and reporting	
	Continual service improvement	

Appendix 3
Some Relevant Standards

Substation communications and environment		
IEC	60834-1	Teleprotection equipment for power systems—performance and testing—command systems
IEC	60870-5-101	Serial communication for SCADA RTU
IEC	60870-5-104	TCP/IP communication for SCADA RTU
IEC	60870-5-6	Telecontrol application service element (TASE-2) protocol
IEC	61850	Communication networks and systems for power utility automation
IEC	61850-90-1	Use of IEC61850 for the communication between substations
IEC	61850-90-2	Use of IEC61850 for the communications between control centers and substations
IEC	61850-90-3	Use of IEC61850 for condition monitoring
IEC	61850-90-4	IEC61850—network engineering guidelines
IEC	61850-90-5	Use of IEC61850 to transmit synchrophasor information
IEC	61850-90-6	Use of IEC61850 for distribution feeder automation systems
IEC	61850-7-420	Communication systems for distributed energy resources (DER)
IEC	62351-6	Security in IEC61850
IEEE	C37.118	Synchrophasor information structure
IEEE	C37.94	Direct fiber interface between the protection and the TDM system
IEC	61000-6-5	Electromagnetic compatibility—generic standards—immunity for power station and substation environments
IEC	61000-4-2	Electrostatic discharge
IEC	61000-4-3	Radiated radio frequency EM field from 80 to 3000 MHz
IEC	61000-4-4	Fast transient/burst
IEC	61000-4-5	Surge 1.2/50 μs line to ground and line to line
IEC	61000-4-6	Conducted disturbance induced by radio frequency fields
IEC	61000-4-11	Voltage dips and voltage interruptions
IEC	61000-4-12	Damped oscillatory wave, common mode, and differential mode
IEEE	1613	Environmental and testing requirements for communication network devices in the electrical power substation
TDM networks		
ITU-T	G.811	Timing characteristics of primary reference clocks (PRC)
ITU-T	G.812	Timing requirements of slave clocks for use as node clocks in synchronization networks

(continued)

© Springer International Publishing Switzerland 2017

C. Samitier (ed.), *Utility Communication Networks and Services*,
CIGRE Green Books, DOI 10.1007/978-3-319-40283-3

(continued)

ITU-T	G.821	TDM error performance
ITU-T	G.823	Control of jitter and wander within digital networks
ITU-T	G.825	Control of jitter and wander within digital networks based on SDH
ITU-T	G.703	Characteristics of hierarchical digital interfaces (E1, PDH)
ITU-T	G.774.x	Synchronous digital hierarchy (SDH)
ITU-T	G.774.4	SDH—subnetwork connection protection (SNCP)
ITU-T	G.7041	Ethernet over SDH—generic framing procedure (GFP) (also Y.1303)
ITU-T	G.7042	Ethernet over SDH—link capacity adjustment scheme (LCAS)
ITU-T	G.7043	Ethernet over SDH—virtual concatenation (VCAT)
Packet-switched Network		
IEEE	802.1ag	Ethernet—connectivity fault management (also ITU-T Y.1731 OAM)
IEEE	802.1aq	Ethernet—shortest path bridging (replaces 802.1w and 802.1D rapid/normal spanning tree)
IEEE	802.1AX	Ethernet—link aggregation control (previously 802.3ad)
IEEE	802.1p	Ethernet priority assignment
IEEE	802.1Q	Ethernet virtual LAN
IEEE	802.1w	Ethernet—rapid spanning tree protocol
IEEE	802.1x	Ethernet security—port-based network access control
IEEE	802.3	Ethernet standards
IEEE	1588v2	Precision time protocol
ITU-T	G.8012.1	Interfaces for the Ethernet transport network
ITU-T	G.8013	OAM functions and mechanisms for Ethernet-based networks
ITU-T	G.8031	Ethernet linear protection switching
ITU-T	G.8032	Ethernet ring protection switching
ITU-T	G.8011/8011.x	Carrier Ethernet services (E-Line, E-LAN, EPL/EVPL, EPLAN/EVPLAN)
ITU-T	G.8260	Definitions and terminology for synchronization in packet networks
ITU-T	G.8261	Timing and synchronization aspects in packet networks
ITU-T	G.8261.1	Packet delay variation network limits applicable to packet-based methods (frequency synchronization)
ITU-T	G.8262	Synchronous Ethernet
ITU-T	G.8264	Distribution of timing information through packet networks
ITU-T	Y.1541	Network performance objectives for IP-based services
ITU-T	Y.1730	Requirements for OAM (operation, administration and maintenance) functions in Ethernet-based networks
ITU-T	Y.1731	OAM functions and mechanisms for Ethernet-based networks
ITU-T	G.8101	Terms and definitions for MPLS transport profile (MPLS-TP)
ITU-T	G.8110	Architecture of MPLS-TP network
ITU-T	G.8112	Interfaces for MPLS-TP
ITU-T	G.8113-1/-2	OAM mechanisms for MPLS-TP
ITU-T	G.872	Architecture of optical transport networks (OTN)

(continued)

(continued)

ITU-T	G.709	Interfaces for the optical transport network (OTN)
IETF	RFC2030	Simple network time protocol (SNTP)
IETF	RFC 1157	Simple network management protocol (SNMPv1)
IETF	RFC 3411/3418	Simple network management protocol (SNMPv3)
IETF	RFC4553	Structure-agnostic time division multiplexing over packet (SAToP)
Physical layer interfaces		
IEEE	802.3ab	1000BASE-T gigabit Ethernet over copper wire
IEEE	802.3z	1000BASE-X gigabit Ethernet over fiber
IEEE	802.3af/at	Ethernet—power over Ethernet (15 W/25 W)
IEEE	802.3av	10 Gbps Ethernet passive optical network (EPON)
IEEE	802.3u	Ethernet—100BASE-TX Fast Ethernet over copper wire
IEEE	802.3bm/ba	Ethernet—40G/100G Ethernet interfaces
IEEE	802.3ac	Ethernet—10G over fiber
Access network technologies		
IEEE	802.11n/ac	Wireless LAN at 2, 4 and 5 GHz (WiFi)
IEEE	802.15.4	Low-rate wireless personal area network (used by ZigBee and 6LoWPAN)
IEEE	802.16/e	Wireless MAN for broadband wireless access
ITU-T	G.991.x	HDSL/SHDSL (high bitrate digital subscriber line)
ITU-T	G.992.x	ADSL (asymmetric digital subscriber line)
ITU-T	G.993.x	VDSL2 (very high bit rate digital subscriber line)
ITU-T	G.998.x	xDSL Ethernet
ITU-T	G.9903–9904	Narrow-band OFDM power line communications (G3-PLC, PRIME)
IEEE	802.11n/ac	Wireless LAN at 2, 4 and 5 GHz (WiFi)
IEEE	802.15.4	Low-rate wireless personal area network (used by ZigBee and 6LoWPAN)
IEEE	802.16/e	Wireless MAN for broadband wireless access
IEEE	1901-1	Low frequency (<500 kHz) Power line communication (PRIME, G3-PLC, etc.)
IEEE	1901-2	Broadband power line communication
Optical communications		
ITU-T	G.651.1	Multimode graded index 50/125 μm fiber for the optical access network
ITU-T	G.652	Characteristics of single mode fibers
ITU-T	G.653	Characteristics of dispersion-shifted single mode optical fibers
ITU-T	G.655	Characteristics of nonzero dispersion-shifted single mode optical fibers
ITU-T	G.671	Coarse wavelength division multiplexing (CWDM)
ITU-T	G.694.1	Dense wavelength division multiplexing (DWDM)
ITU-T	G.984.x	Gigabit-capable passive optical networks (G-PON)

Appendix 4
CIGRE Technical Brochure Contributors

Ethernet in power utility TB460 (WGD2.23)		Telecom service in the power utility TB461 (WGD2.26)	
C. Samitier (Conv.) (Spain) M. Mesbah (Sec.) (France)		M. Mesbah (conv.) (France)	
J. Fonseca (Portugal)	O. Finnekaasa (Norway)	R. Evans (Australia)	J.A. Garcia Lopez (Spain)
P. Cooney (Ireland)	C. Huntley (Canada)	D. Bell (Australia)	S. Javornik Voncina (Croatia)
A. Runesson (Sweden)	R. Pellizzoni (Argentina)	J. Piotrowski (Poland)	J. Leifer (Israel)
A. Cadenas (Spain)	C. Trigo de Loureiro (Brazil)	J. Mendes (Portugal)	M. Yamasaki (Japan)
A. Arzuaga (Spain)	D. Bell (Australia)	P. Gama (Portugal)	K. Iwasaki (Japan)
E. Marquesini (United Kingdom)	J. Piotrowski (Poland)	I. Nedelcu (Romania)	E. Lassner (Hungary)
P. Schwyter (Switzerland)	M. Inoue (Japan)	M. Blokar (Slovenia)	L. Lhassani (Netherland)
S. Yliraasakka (Finland)	J. Gajica (Serbia)	P. Schwyter (Switzerland)	O. Stokke (Norway)
R. Elliott (South Africa)	W. Azlan (Malaysia)	P. Renshall (United Kingdom)	J. Fonseca (Portugal)
		C. Trigo de Loureiro (Brazil)	P. Marques (Portugal)
		E. Bandeira de Melo (Brazil)	D. Lalovic (Serbia)

© Springer International Publishing Switzerland 2017

C. Samitier (ed.), *Utility Communication Networks and Services*,
CIGRE Green Books, DOI 10.1007/978-3-319-40283-3

Operation and maintenance of telecom networks TB588 (WGD2.33)		Line and system protection communications TB521 (JWGD2.B5.30)	
M. Mesbah (conv.) (France) P. Parisot (Sec.) (France)		M. Mesbah, C. Samitier (Conv.) (France, Spain) T. Einarsson (Sec.) (Sweden)	
Th. Leroy (Belgium)	I. Nedelcu (Romania)	M. Acacia (Belgium)	J. Freitas (Portugal)
J. Piotrowski (Poland)	J. Crisp (New Zealand)	J. Alvarez (Spain)	A. Runesson (SWEDEN)
A. Runesson (Sweden)	F. Castro (Spain)	H. Spiess (Switzerland)	G. Vianello (Italy)
J. Alvarez (Spain)	D. Lalovic (Serbia)	M. Valente (Italy)	C. Komatsu (Japan)
J. Darne (Spain)	R. Gomez (Argentina)	Th. Leroy (Belgium)	M. Stockton (United Kingdom)
P. Renshall (United Kingdom)	R. Fernandez (Australia)	U. Carmo (Brazil)	I. Viziteu (Romania)
R. Francisco (Portugal)	M. Scott (Australia)	F. Castro (Spain)	M.A. Ordunez (Spain)
K. Iwasaki (Japan)	K. Darwish (Arab Emirates)	J. Darne (Spain)	A. Struecker (Germany)
Z. Gydien (South Africa)		R. Cimadevilla (Spain)	J. Wright (United Kingdom)
		S. Dollerup (Denmark)	

Bibliography

Telecommunication Service Provisioning and Delivery in the Electrical Power Utility: CIGRE Technical Brochure 461, Working Group D2.26 (April 2011)

The use of Ethernet Technology in the Power Utility Environment: CIGRE Technical Brochure 460, Working Group D2.23 (April 2011)

Line and System Protection using Digital Circuit and Packet Communications: CIGRE Technical Brochure 521, Working Group D2/B5.30 (December 2012)

Operation and Maintenance of Telecom Networks and Associated Information Systems in the Electrical Power Utility: CIGRE Technical Brochure 588, Working Group D2.33 (July 2014)

Building IP Capability in the Operational Telecom Network to leverage Smart Applications in the Power Grid, M. Mesbah, Alstom Grid White Paper

Packet-switched Communications for Utility Operational Telecom Networks, M. Mesbah, GE Energy Connection White Paper

An Integrated Management Solution for Operational Telecom Networks in Electrical Power Utilities, M. Mesbah, Alstom Grid White Paper

Communication Architecture for IP-based Substation Applications: CIGRE Technical Brochure 507, Working Group D2.28 (August 2012)

Scalable Communication Transport Solutions over Optical Networks: CIGRE Technical Brochure 618, Working Group D2.35 (May 2015)

© Springer International Publishing Switzerland 2017

C. Samitier (ed.), *Utility Communication Networks and Services*,
CIGRE Green Books, DOI 10.1007/978-3-319-40283-3

Index

© Springer International Publishing Switzerland 2017
C. Samitier (ed.), *Utility Communication Networks and Services*,
CIGRE Green Books, DOI 10.1007/978-3-319-40283-3

FSC
www.fsc.org
MIX
Papier aus verantwortungsvollen Quellen
Paper from responsible sources
FSC® C105338

If you have any concerns about our products, you can contact us on
ProductSafety@springernature.com

In case Publisher is established outside the EU, the EU authorized representative is:
Springer Nature Customer Service Center GmbH
Europaplatz 3, 69115 Heidelberg, Germany

Printed by Libri Plureos GmbH
in Hamburg, Germany